Hope and Courage in the Climate Crisis

"A vitally important book."
—Prof. Tim Flannery, *author of* The Future Eaters (1994) *and* The Climate Cure (2020)

"A great book to guide us into the depths of who we really are and out of the overwhelm of the climate crisis."
—Christiana Figueres, *co-author of* The Future we Choose *and Executive Secretary of the UNFCCC (2010–2016)*

"This book focuses on some of the most creative and courageous responses to the crisis that hangs over all our lives: these are stories that need to be told and retold, so that we know we are not alone in facing this monumental challenge."
—Bill McKibben, *Founder of 350.org*

John Wiseman

Hope and Courage in the Climate Crisis

Wisdom and Action in the Long Emergency

John Wiseman
Melbourne Sustainable Society Institute
University of Melbourne
Parkville, VIC, Australia

ISBN 978-3-030-70742-2   ISBN 978-3-030-70743-9 (eBook)
https://doi.org/10.1007/978-3-030-70743-9

© The Editor(s) (if applicable) and The Author(s), under exclusive license to Springer Nature Switzerland AG 2021
This work is subject to copyright. All rights are solely and exclusively licensed by the Publisher, whether the whole or part of the material is concerned, specifically the rights of translation, reprinting, reuse of illustrations, recitation, broadcasting, reproduction on microfilms or in any other physical way, and transmission or information storage and retrieval, electronic adaptation, computer software, or by similar or dissimilar methodology now known or hereafter developed.
The use of general descriptive names, registered names, trademarks, service marks, etc. in this publication does not imply, even in the absence of a specific statement, that such names are exempt from the relevant protective laws and regulations and therefore free for general use.
The publisher, the authors and the editors are safe to assume that the advice and information in this book are believed to be true and accurate at the date of publication. Neither the publisher nor the authors or the editors give a warranty, expressed or implied, with respect to the material contained herein or for any errors or omissions that may have been made. The publisher remains neutral with regard to jurisdictional claims in published maps and institutional affiliations.

Cover credit: enjoynz/gettyimages

This Palgrave Macmillan imprint is published by the registered company Springer Nature Switzerland AG
The registered company address is: Gewerbestrasse 11, 6330 Cham, Switzerland

Acknowledgements

Guy Abrahams, Sam Alexander, Ammar Aldaoud, Stephen Ames, Ian Barnes, Jeremy Baskin, Avril Blay, Grant Blashki, Susie Burke, Fay Chomley, Gerald Frape, Dee Gill, Gerry Gill, Elly Harrould-Kolieb, David Karoly, Victoria McKenzie-McHarg, Monica Minnegal, Stephen Pollard, Carol Ride, Daniel Simons, David Spratt, Sebastian Thomas, Andrew Wiseman, Daniel Wiseman, Jemma Wiseman, Annabelle Workman.

Contents

1	**Facing a Harsh Climate Future with Eyes Wide Open**	1
	The Necessity and Urgency of Emergency Speed Climate Action	3
	Meaning and Purpose in a Burning World	5
	Defiant Hope and Radical Courage	7
	Overview of Themes and Chapters	11
2	**Beyond Denial and Despair: Honesty and Action on the Climate Change Front Line**	19
	Tell Them About the Water: Climate Change as an Existential Risk	21
	Beyond Denial: The Scientific and Ethical Case for Swift and Decisive Climate Action	27
	Pathways to a Just and Resilient Zero-Carbon Future	30
	Beyond Despair: Wisdom and Resilience in the Long Emergency	33
3	**Remembering Magnificence: Collective Action and the Beauty of the Earth**	37
	Contemplating the Beauty of the Earth: Rachel Carson and *Silent Spring*	38

	The Only Limitations We Have Are Our Own Imaginations: *350.org*	41
	We Are Not Drowning, We Are Fighting! *the Pacific Climate Warriors*	44
	We Did This Together. We Did This With Art. We Did This As Art *Liberate Tate*	47
	We Need Social Tipping Points: *The German Energiewende*	50
	The Great Society, the Moon Shot, the Civil Rights Movement of Our Generation: The *Sunrise Movement* and the *Green New Deal*	54
	I Want You to Act Is if Your House Is on Fire. Because It Is: *School Strike for Climate*	56
4	**Caring for Country: Indigenous and First Nations Learning About Survival, Resilience and Resistance**	**61**
	If We Fail to Care for Country, It Cannot Care for Us	63
	If You Don't Move with the Land, the Land Will Move You	66
	Climate Justice at Emergency Speed	69
	We Are Showered Every Day with the Gifts of the Earth	73
	Overcoming the Windingo: Attentiveness, Gratitude, Reciprocity and Healing	75
5	**Cooling the Fevered City: Reason and Hubris in Greek and Enlightenment Philosophy**	**77**
	Critical Analysis and Rigorous Debate	81
	The Just Society and the Virtues of Moderation	83
	Making the Best of What Is in Our Power	87
	Have Courage to Use Your Own Reason!	88
	Ethical Action in an Interconnected Universe	92
	Flesh, Blood and Brain, We Belong to Nature	95
6	**Illuminating the Patterns of Domination: Critical Theory, Modernity and Power**	**99**
	Illuminating the Patterns of Domination	101
	Reconnecting World and Earth	103

	The Incalculable Grace of Love	105
	Respectful and Well-Informed Deliberative Decision Making	107
	Creativity and Beauty, Love and Friendship	111
	The Responsibilities of Freedom	113
	Confronting the Violence of Othering	115
	Mending What Has Been Torn Apart	117
	Join up, Protest, Propose, Create	119
7	**Mercy to All Beings: Learning from Christian, Jewish and Islamic Traditions About Thankfulness, Love and Care**	**123**
	The Cry of the Earth and the Cry of the Poor: Christianity, Ecology and Climate Change	125
	The Smoke of Grief and the Fire of Love	128
	Faith and Evidence; Suffering and Care	130
	Between the Fires: Judaism, Ecology and Climate Change	133
	We Will Again Stand Mountain-Strong: *Tisha B'Av* in a Time of Climate Crisis	135
	The World Is Sweet and Verdant: Islam, Ecology and Climate Change	137
	The Islamic Declaration on Global Climate Change	140
	Sometimes We Are Compelled to Act: Faith-Based Engagement with Extinction Rebellion	141
8	**This World Is but a Dew Drop World....and yet....: Buddhist, Taoist and Confucian Learning About Suffering, Impermanence and Compassion**	**145**
	Sitting Still and Sweeping the Garden: Insights and Pathways of Engaged Buddhism	146
	People Who Love the Earth: Remembering Why We Care	149
	Overcoming Ignorance, Violence and Greed	151
	Interdependence and Impermanence; Compassion and Generosity	153

	People Follow Earth, Earth Follows Heaven, Heaven Follows the Way: Taoist Wisdom in a Time of Climate Crisis	159
	All Under the Heavens Belongs to the Public: Ecological Intelligence in Confucian Thought	162
9	**Living Ecologically: Understanding and Respecting Complexity and Fragility**	**167**
	So Delicately Interwoven Are the Relationships…	168
	Pathways to an Ecology of Mind	171
	There Is Such a Thing as Enough	174
	Shallow and Deep Ecology	178
	We Are Not the Centre	181
	Where Should We Land?	185
	Homecoming to the Earth	190
10	**How the Light Gets In: Imagining and Creating Just and Resilient Zero-Carbon Worlds**	**195**
	Other Worlds Are Possible	196
	Imagining the Bridge: Disruptive Transformation in a Post-pandemic World	200
	Creating a Just and Resilient Zero-Carbon Future	202
	Justice and Care for All Creation	204
	Creating a Practice of Resilience, Replenishment and Regeneration	206
11	**The 2050 Zero-Carbon World Oration**	**211**
	The Age of Foolishness or the Age of Wisdom? The View from 2020	213
	Milestones on the Journey to a Zero-Carbon World	216
	Drivers of the Great Energy Transition	216
	The Escalating Frequency and Severity of Catastrophic Climatic Events	220
	Disruptive, Game-Changing Technological Innovation	221
	Disruptive Game-Changing Innovation in Social, Economic and Political Systems	222
	A Great Leap Forward on a Long and Challenging Road	224

12 The Laughter of Children, the Roar of the Ocean:
    Concluding Thoughts and Questions   227

Appendix A: Donella Meadows, Leverage Points—Places
             to Intervene in a System   233

Appendix B: Bali Principles of Climate Justice, 2002   235

Appendix C: Universal Declaration of Rights of Mother
             Earth, World People's Conference
             on Climate Change and the Rights
             of Mother Earth, Cochabamba, Bolivia
             April 2010   239

References   245
Index   271

1

Facing a Harsh Climate Future with Eyes Wide Open

In whatever time and place you are reading these words, I wonder how your life and your world are unfolding? How fierce is the heat of your days and how wild are your storms? What colour is the sky above you? What are the hopes and fears and dreams of your family, your friends and your community?

On 20 September 2019, I joined over one hundred thousand climate strike demonstrators marching through the streets of Melbourne. Like many others, I was powerfully inspired by the passion and determination of millions of young people all around the world rising up and demanding climate action at the speed and scale required to keep global warming below 1.5 degrees. A few days later I was also one of millions who heard Greta Thunberg's searing condemnation of world leaders for their continuing failure to commit to decisive climate action.

> We are in the beginning of a mass extinction, and all you can talk about is money and fairy tales of eternal economic growth. How dare you! For more than 30 years, the science has been crystal clear. How dare you continue to look away and come here saying that you're doing enough, when the politics and solutions needed are still nowhere in sight. [1]

Seven years earlier, in 2012 I had the honour of interviewing many of the world's leading climate scientists, activists and policy makers about plans and strategies for rapidly accelerating the transition to a post-carbon economy [2]. Their conclusions were clear and consistent. Catastrophic climate change could only be avoided if the level of greenhouse gasses in the atmosphere began to fall immediately. The actions required to achieve this goal were all technologically achievable and financially affordable. The most urgent task therefore was to rapidly remove the political roadblocks preventing the swiftest possible acceleration of climate action. There was also broad agreement that real progress towards achieving these goals would need to be well underway by 2015.

Now, in 2021, at the time of writing, we face an even more confronting reality. Despite the enormous hard work and commitment of many millions of people, the political obstacles preventing decisive climate action remain formidable and greenhouse gas levels continue to rise. The climate science evidence becomes more confronting every day. Images of increasingly severe, increasingly frequent fires, storms and floods continue to fill our screens.

My decision to write this book has been informed by many hundreds of conversations about climate change, hope and courage with students and scientists; artists and activists; children and grandparents; farmers and film-makers—all struggling to find ways of living creative, meaningful lives on a rapidly heating, increasingly unfamiliar planet. These conversations very often include observations such as the following. 'I completely understand and support the need for decisive action to reduce greenhouse gas emissions. But I also know that we are in for a very rough ride no matter how fast we act. There are times therefore, perhaps at four in the morning—or when picking up my grandchild—or when thinking about the full consequences of many metres of sea level rise—when I struggle to avoid despair.'

Human beings have found many creative and inspiring ways to endure and even thrive in dark and threatening times. An abiding, foundational belief underpinning many of these insights has been the confidence that, in the end, the 'great wheel of history' will turn again and that Martin Luther King was correct in his reassuring assertion that 'the arc of history bends towards justice.' Three key differences however characterise the era of rapidly accelerating global warming in which we are now living.

First, the existential climate change risks we are now facing are not confined to particular communities or societies but to the overall fabric of the Earth's ecologies and ecosystems which have provided a fertile home for the evolution of human society—and of millions of other species—over the last 10,000 years.

Second, the threats created by potentially catastrophic global warming have, to a large extent, been created by the choices and actions of human beings, with by far the greatest responsibility resting with the most affluent and powerful.

Third, while the complete disappearance of human beings from the Earth may be unlikely in the foreseeable future, global warming beyond 3 or 4 degrees has the potential to remove the foundational conditions required to sustain civilised human societies. Climate change risks and impacts are also further intensifying and compounding a wide range of other ecological threats and tipping points. We all live with the growing realisation that this is likely to be a very long emergency.

The Necessity and Urgency of Emergency Speed Climate Action

As the risks of catastrophic climate change continue to grow, so too do the challenges of facing an increasingly hazardous future with honesty and courage; justice and compassion; meaning and purpose. This book therefore aims to bring together and explore diverse sources of wisdom and insight which can strengthen our capacity to live courageous, compassionate and creative lives in a world of rapidly accelerating climatic and ecological risk.

Evidence from disaster response researchers reminds us that effective and timely responses to fast-moving fires, floods and storms depend on recognition that swift, decisive action is necessary and urgent (the emergency is real and heading our way); possible within the time available (there is a clear course of action which will significantly reduce the danger); and desirable (the benefits of action clearly outweigh the risks and dangers of inaction) [3].

While future readers may find this hard to understand, climate change denial was still a potent force in countries like Australia and the United States in 2020. The vast majority of the world's citizens along with many millions of climate scientists, military leaders, investment analysts, health professionals and emergency service workers are now however fully convinced about the scientific, ethical and financial case for decisive, rapid action to reduce and address climate risk.

The scientific and experiential evidence of the accelerating, existential risks of climate change is abundantly clear. As greenhouse gas emissions continue to rise heatwave and extreme weather records around the world continue to be broken. The escalating ferocity of cyclones sweeping over the Philippines, floodwaters surging again in Bangladesh and the severity of droughts in sub-Saharan Africa all add to mounting evidence that climate change consequences are likely to be greatest for the most vulnerable individuals and communities.

Endorsement of long-term global warming goals of 1.5 °C and the achievement of net zero emissions by the second half of the century were welcome outcomes of the 2015 Paris Climate Summit. The abiding problem, as the world's climate scientists continue to note, is that current (2020) national climate policy commitments are in fact leading the world into the existential risk zone of global warming well beyond 3 or 4 °C [4].

The timetable and actions required to achieve emission reductions at the speed and scale required to significantly reduce climate risks are also well understood. There is now broad scientific agreement that keeping global temperatures below 1.5°C will require greenhouse gas emissions to peak by the early 2020s followed by rapid reduction to zero by as close as possible to 2030 [4]. Longer timetables for achieving zero emissions will require stronger action to draw down CO_2.

The actions required to achieve these goals are well known. Lowering energy demand by reducing consumption and improving energy efficiency; accelerating the transition from fossil fuels to renewable energy; drawing down CO_2 emissions in ways consistent with social equity and the health and well-being of all communities and ecosystems; expanding investment in climate adaptation and resilience; and prioritising actions to reduce climate impacts for the most vulnerable individuals and communities. While these actions do indeed require a rapid shift in global financial, economic and social priorities and resources, the opportunities and benefits from decisive action far outweigh the costs.

The greatest barriers to accelerating the action required to drive an emergency speed transition to a zero-carbon economy are social and political—not technological or financial. The largest obstacles which need to be overcome are individual and collective denial of the speed with which action needs to be taken combined with the political power and influence of those who gain the most from maintaining a fossil fuel-based economy. Courageous leadership, creativity and collective action to overcome these barriers and accelerate emission reduction remain the most crucial and urgent political priorities.

Meaning and Purpose in a Burning World

Noting the global warming trends already in the pipeline, most plausible climatic scenarios still also require us to imagine and prepare for a planetary climate far harsher than the relatively benign Holocene world which human beings have been fortunate enough to inhabit for the last 10,000 years. While climate change is arguably the greatest long-term risk other rapidly escalating ecological threats including the extinction of species, the destruction of forests and the acidification of oceans are all of deepening concern.

Securing the political support to drive emergency speed climate action is also becoming harder in the context of increasing inequality, insecurity, racism and xenophobia combined with the growing success of right-wing and populist political parties and politicians. The sudden arrival of COVID-19 has provided further evidence of the many ways in which differing responses to global existential risks such as a pandemic can reinforce or reduce glaring inequalities of health and income; wealth and power. The COVID-19 pandemic may lead to some short-term reductions in emissions. There are also significant risks that the impact of the pandemic will divert attention and resources away from climate crisis challenges and solutions.

A great deal of hard work and a lot of luck may enable us to avoid some of the most dangerous climate change and ecological tipping points. Current and future generations are however on a journey into a world of more frequent and severe extreme weather events; more heatwaves, fires, floods and famines; and more rapid extinctions of animals, birds and insects. This journey will be particularly hard for the poorest and most vulnerable of peoples.

Awareness of the likelihood that even the most rapid and decisive action to decarbonise the global economy will still be insufficient to prevent severe and irreversible social and ecological impacts is a source of deep concern and distress for many individuals passionately committed to decisive climate action. The second tough and urgent task therefore—and the primary focus of this book—is to address the question: What sources of wisdom and insight can strengthen our capacity to take courageous and effective action and to live meaningful and creative lives in a world of rapidly accelerating climatic and ecological risks?

My aim in writing *Hope and Courage in the Climate Crisis* is to contribute to this conversation by exploring ways in which wisdom and learning from diverse traditions: from climate scientists and activists; psychologists, philosophers and social theorists; Indigenous cultures and ways of life; faith-based and spiritual perspectives; artists and writers can assist us face a harsh climate future with honesty and courage, meaning and purpose.

This short book does not attempt to duplicate the many excellent guides to the emotional resilience, self-care skills and daily practices which can help us cope with the increasingly distressing psychological consequences of climate grief and eco-anxiety [5]. *Hope and Courage in the Climate Crisis* is rather best understood as a complementary series of reflections from my ongoing journey through the libraries and galleries; stories and conversations; ideas and debates which people I know and respect have found to be valuable sources of radical hope and defiant courage in threatening times.

These diverse reflections, ideas and insights can in my experience provide valuable conceptual frameworks and analytic tools enabling us to more clearly identify the size and speed of the approaching storm; the forces driving the tempest towards us and the actions we can take to deal with the most dangerous and immediate threats. In the longer term, these insights and ideas can also assist us design and create alternative ways of thinking and acting which can help us navigate the wild and alien landscapes of the long emergency.

Defiant Hope and Radical Courage

Before outlining the key themes and structure of the book, it is important to be clear about the ways in which I am using the words 'hope' and 'courage.' The language of hope has always attracted passionate supporters as well as fierce critics. For some, like radical Welsh author and activist Raymond Williams, an appropriately well informed, well-grounded sense of hope is fundamental to effective political action: 'to be truly radical is to make hope possible, rather than despair convincing' [6, p. 118]. For others, like German philosopher Friedrich Nietzsche, the language of hope is fatally infected by the superficially attractive but profoundly dangerous diseases of wishful thinking and false optimism: 'Hope is the worst of all evils because it prolongs the torments of man' [7, p. 71].

Norwegian psychologist, Per Espen Stoknes provides the following useful typology of the various ways in which 'hope' is used to inform responses to climate risk and climate action [8].

> Passive optimism or 'Pollyanna Hope,' in which a person believes in a positive (e.g., safe, bright, thriving) future that will simply come about on its own, or by someone else's doing (e.g. god, nature, or some technological fix).
> Active optimism or 'Heroic Hope,' in which the person has a similarly positive outlook but understands he or she needs to actively help bring it about.
> Passive scepticism or 'Stoic Hope,' in which a person is not at all convinced that the future will be bright and easy, but believes not much needs to be done because it will be bearable.

Active scepticism or 'Grounded Hope,' in which a person is realistically informed about the state of affairs, and thus sceptical of a positive outlook, but chooses to do whatever she or he can to bring about decisive action.

Concern about the false comfort and complacency arising from naively optimistic 'hopefulness' leads ecological author and activist Derrick Jensen to the following cautionary and provocative conclusion, 'frankly, I don't have much hope. But I think that's a good thing.'

> Hope is what keeps us chained to the system, the conglomerate of people and ideas and ideals that is causing the destruction of the Earth....Hope is, in fact, a curse, a bane. I say this not only because of the lovely Buddhist saying 'Hope and fear chase each other's tails,' not only because hope leads us away from the present, away from who and where we are right now and toward some imaginary future state. I say this because of what hope is. Hope is a longing for a future condition over which you have no agency; it means you are essentially powerless [9].

Other equally passionate and committed climate activists including Rebecca Solnit, Susan Moser and Joanna Macy argue that the more robust framing of 'grounded,' 'active' or 'defiant' hope can still usefully illuminate the necessity and urgency of ethically and scientifically informed agency and engagement. For Rebecca Solnit, author of *Hope in the Dark: Untold Histories, Wild Possibilities* hope is 'an act of defiance, or rather the foundation for an ongoing series of acts of defiance, those acts necessary to bring about some of what we hope for and to live by principle in the meantime' [10, p. 163]. 'Hope' Solnit adds 'is not about what we expect. It's an embrace of the essential unknowability of the world....hope is not a door but a sense that there might be a door' [11].

For climate resilience and social transformation researcher and adviser, Susan Moser the concepts of grounded, active or authentic hope refer to the situation 'where you are not at all convinced that there is a positive outcome at the end of your labors. It's not like you're working towards winning something grand. You don't know that you'll able to achieve that. But you do know that you cannot live with yourself if you do not do everything toward a positive outcome' [12].

Active hope, for Buddhist scholar and activist Joanna Macy involves 'a readiness to discover the size and strength of our hearts, our quickness of mind, our steadiness of purpose, our own authority, our love for life, the liveliness of our curiosity, the unsuspected deep well of patience and diligence, the keenness of our senses, and our capacity to lead' [13, p. 35].

The concept of 'radical hope,' as explored by American philosopher and psychoanalyst Jonathon Lear starts from a bleaker assessment of the inevitability of species extinctions, degraded ecosystems and human suffering. Lear, drawing on the experience of Plenty Coups, last chief of the Crow Nation who led his people through a period of profound cultural devastation, argues that we should choose to act with moral integrity and courage even if we understand full well that we are facing a time of mass extinctions and severe ecological degradation. Radical hope, Lear argues 'is against despair, even in the face of a well-justified despair. It is the idea that an inadequate grasp of the good should not lead one to believe it is not to be hoped for' [14, p. 49].

Despite all of the important work undertaken to infuse the concept of 'hope' with a stronger sense of defiance and activism, 'hope' still retains for most people many of the more limited attributes of the Oxford Dictionary definition: 'A feeling of expectation and desire for a particular thing to happen; grounds for believing that something good may happen; aspiration, desire, wish, dream, longing, yearning, craving, daydream, pipe dream.....' [15].

Courage, understood as 'the ability to do something that frightens one, of bravery, daring, audacity, boldness, backbone, fortitude, and resolution' may therefore be a more helpful description of the qualities required to nurture and sustain human meaning and purpose in the context of increasingly daunting, existential threats of climate catastrophe.

Courage for Nelson Mandela is 'not the absence of fear, but the triumph over it. The brave man is not he who does not feel afraid, but he who conquers that fear' [16]. Courage, for psychologist Rollo May, 'is not the absence of despair; it is, rather, the capacity to move ahead in spite of despair' [17, p. 12]. 'Courage,' the poet Maya Angelou argues 'is the most important of all the virtues. Because without courage, you cannot practice any other virtue consistently. You can be kind for a while; you can be generous for a while; you can be just for a while, or merciful for a while, even loving for a while. But it is only with courage that you can be persistently and insistently kind and generous and fair' [18].

NASA climate scientist Kate Marvel draws on all these varying interpretations of courage in her reflections on the skills and capabilities we need to learn to acknowledge, mourn and rise above our climate grief and fear.

> The scale of climate change engulfs even the most fortunate. There is now no weather we haven't touched, no wilderness immune from our encroaching pressure. The world we once knew is never coming back. I have no hope that these changes can be reversed. We are inevitably sending our children to live on an unfamiliar planet. But the opposite of hope is not despair. It is grief. Even while resolving to limit the damage, we can mourn. And here, the sheer scale of the problem provides a perverse comfort: we are in this together. The swiftness of the change, its scale and inevitability, binds us into one, broken hearts trapped together under a warming atmosphere.
>
> We need courage, not hope. Grief, after all, is the cost of being alive. We are all fated to live lives shot through with sadness and are not worth less for it. Courage is the resolve to do well without the assurance of a happy ending [19].

The experience and observations of scientists and activists; health professionals and psychologists; firefighters and emergency service workers on the climate change front line strengthen our understanding that the first, crucial step in dealing constructively with trauma and grief is to face the harsh realities of escalating climate risk honestly and with our eyes wide open. We can also learn a great deal from these experiences about the diversity of insights and resources which individuals and communities from differing backgrounds and traditions draw on to overcome despair; sustain emotional resilience and inspire creative and courageous action.

Some of us, for example, draw strength and wisdom from the shared visions, values and companionship of collective political action. Some are inspired and sustained by Western philosophical insights about the potential for learning and reason to drive the ethically informed creativity and inventiveness required to meet great social and ecological challenges. Others find comfort in the teachings of Indian and Chinese philosophies about impermanence and interdependence or from Christian, Jewish and Islamic faith-based traditions about the importance of sustaining love and compassion in the face of suffering. Others again turn to art and literature; music and film to remind us that human beings are capable of acting with courage and compassion as well as with cruelty and violence and to assist us to imagine and visualise the kind of world we hope to create and the pathways that can lead us there.

Overview of Themes and Chapters

Chapter 2, *Beyond denial and despair* begins by reaffirming the importance of an honest appraisal of climate change risks and consequences as an essential foundation for effective action. Closing our eyes to the approaching storm, wishing the storm would go away or becoming paralysed by despair are three different but equally unhelpful recipes for disaster. In an age of competing and contested claims about truth and consequences, it is valuable to pay attention to evidence from a variety of sources. The scientific evidence about climate change trends and causes, risks and implications is compelling. But so too is the lived experience of Indigenous communities; farmers and firefighters; artists and writers; health and emergency service workers on the climate change front line.

The psychological and political forces driving denial of the evidence of anthropogenic climate change are now well understood. The drivers of a second and perhaps even more dangerous form of denial are also increasingly clear: wishful thinking and false optimism leading to denial of the urgency and scale of action required to significantly reduce climate risks. The roadblocks preventing rapid and effective action are also well known: the power and influence of the fossil fuel industry and other vested interests; social and technological path dependencies; financial, governance and implementation constraints; and the dominant neoliberal economic paradigm of unsustainable consumption and inequitable wealth distribution.

The greatest challenge for many climate scientists, activists and citizens is overcoming paralysis resulting from grief and despair in confronting the likely consequences of climate change for human beings, other species and for the Earth's ecosystems. Psychologists, health workers and activists have begun to explore a wide range of strategies for strengthening emotional resilience and for sustaining individual and collective action.

Chapter 3, *Remembering magnificence* honours and celebrates the passion and courage of citizens and social movements committed to removing the political roadblocks standing in the way of a rapid transition to a just and sustainable zero-carbon future. Human history is full of stories of extraordinary bravery and collective action driving transformational change which few at the time saw coming: the abolition of slavery, the triumph of the Suffragettes, the overthrowing of apartheid, the fall of the Berlin Wall, the independence of Timor-Leste.

Few people at the time would, for example, have predicted that the publication in 1958 of *Silent Spring* by American biologist Rachel Carson would play such a critical role in launching and inspiring the modern environmental movement. Like many other climate activists and scientists, I deeply admire Carson's skilful communication of the most rigorous scientific evidence and her fierce commitment to speak truth to power no matter what the cost. I also find great strength in the eloquence and poetry of her writing about the diversity, resilience and splendour of life on Earth.

These themes are explored further through the experience of more recent climate action initiatives including *350.org*, *The Pacific Climate Warriors*, *Liberate Tate*, the German *Energiewende*, the *Sunrise Movement for a Green New Deal* and *School Strike for the Climate*. While these movements and campaigns vary greatly in their influence and impact, they all illustrate the power of collective action in overcoming isolation and despair as well as in triggering the social and political tipping points required to accelerate transformational social change.

Chapter 4, *Caring for Country* shares and reflects on learning and insight from four Australian and American Indigenous and First Nation authors and activists whose work I have found particularly helpful in strengthening my understanding of climate justice and climate action priorities. Their work provides a crucial foundation for non-Indigenous readers in recognising the need to fully acknowledge and decisively address the many ways in which climate change intensifies ongoing legacies of colonial violence and displacement. While very conscious of the risks of romanticising and appropriating Indigenous and First Nation knowledge I continue to learn a great deal from paying close attention to the heightened importance of 'caring for country' during periods of disruptive climatic, ecological and social change.

Australian Indigenous authors and activists Tony Birch and Tyson Yunkaporta note that the wisdom and practice of caring for country have multiple dimensions and implications. Recognising and respecting the complexity and fragility of the environments and ecologies in which we live. Strengthening our awareness that narcissistic choices about the resources we consume and the waste we leave behind are likely to have bitter implications for many other species and for future generations. And listening more carefully to the learning of peoples who have lived on this land for many thousands of years about sustainable and resilient fire and food and forest management practices. In reading the work of American First Nation writers Kyle Whyte and Robin Wall Kimmerer I am struck by many similarities with Australian Indigenous perspectives including their searing critique of settler colonial legacies and the timeliness of their emphasis on rebuilding ways of life informed by trust and reciprocity; thankfulness and healing.

Cooling the fevered city starts with Sir David Attenborough's observation that human beings have a long track record of drawing on reason and ingenuity to enable the species to overcome potentially overwhelming threats. This uniquely human capacity for rational analysis and technological inventiveness led classical Greek philosophers such as Socrates, Plato and Aristotle to emphasise the role of critical analysis, rigorous questioning and respectful debate in addressing complex individual and societal threats and challenges.

Many of the most influential philosophers of the European Enlightenment drew on the Athenian philosophical legacy to build the case for scientifically informed problem-solving as the engine room of civilisational progress. An alternative strand of Enlightenment thought, inspired in particular by the work of Dutch philosopher Baruch Spinoza, has been more cautious in warning of the dangers of an over-reliance on rational analysis and technological mastery.

Marxist analysis of the ecological implications of capitalist economic relations also continues to illuminate ways in which the wealth and power of a small minority have been built on the exploitation of human labour and the Earth's resources. From this perspective, any lasting solution to the climate crisis must also involve a systemic transformation to more just and democratic social and economic relationships.

These sharply different assumptions about the relationship between reason and progress; modernity and power continue to inform ongoing debates between champions of technological solutions and 'ecological modernisation' such as Stephen Pinker and the *Breakthrough Institute* and those such as Paul Kingsnorth and the *Dark Mountain* project who see technological hubris as a primary driver of ecological and climatic crisis.

Sisyphus in Flames considers the contribution which the work of critical social theorists and existential philosophers can make to sustaining resistance and resilience in dark times. German social theorists Theodor Adorno and Max Horkheimer, writing at the end of World War II argue that the highest, ongoing priority is to focus razor-sharp critique on the root causes of humanity's self-destructive tendencies: the pathological myth that we can overcome our fear of the unknown through mastery of nature and the pursuit, at any cost of technocratic knowledge and power.

The decision by Adorno's Frankfurt school colleague, Martin Heidegger to prioritise repair of the interconnections between human beings and the Earth helps explain the ongoing influence which his ideas have had for ecological philosophers and activists. Hannah Arendt draws on her close observation of life and death in Hitler's Germany to interrogate the recurring tendency for pathways to the abyss to be opened as much by the failure of decent human beings to act as by the deliberate malevolence of totalitarian psychopaths. Our capacity to stand together, arm in arm, embracing 'the unpredictable hazards of friendship and sympathy…..the great and incalculable grace of love' is, in contrast Arendt argues our greatest source of emotional, psychological and societal resilience.

Existential philosophers Jean Paul Satre, Simone de Beauvoir and Albert Camus deepen and extend these arguments about freedom and responsibility; love and solidarity by noting that our choices as individual human beings are always made in the context of awareness that no individual exists alone. Genuine freedom and true courage therefore mean choosing a side in ways which enable and strengthen the freedom of others as well as ourselves. This is not an argument for selfless altruism but for recognising that our lives and actions only gain meaning through reciprocal relationships with others.

In *The Myth of Sisyphus* Albert Camus addresses the challenge which he describes as 'the one truly serious philosophical problem.' How do we sustain meaning and purpose in the face of overwhelming evidence that all our dreams and actions will finally end in darkness? Camus reviews a range of possibilities: romantic love and hedonistic passion; the intensity of performance; the thrilling conquest of the adventure; and the creative joyfulness of the artist. While all of these choices have their champions and compensations, Camus argues that they are all, in the end elaborate distractions from the stark reality of our predicament: the only truly authentic response is to fully acknowledge the overwhelming absurdity of human existence. And then to join with others in defiant, fierce and passionate rebellion.

Chapter 7, *Mercy to all beings* explores the contribution of Christian, Jewish, and Islamic spiritual and faith-based traditions to climate emergency responses and strategies. While noting criticisms of some theological traditions for their foundational assumptions about the mastery of human beings over nature, these observations are informed by a diverse array of Christian, Jewish and Islamic teachings about thankfulness, love and care.

The Papal Encyclical, *Care for Our Common Home* foregrounds Saint Francis's teachings about our responsibility to honour and care for all the creatures and wonders of God's creation in advocating an ethic of Earth stewardship and climate justice [20]. Jewish theologians reach a similar conclusion about climate justice priorities and responsibilities from their understanding of God's advice to Adam about celebrating and protecting the beauty of the Earth. The hymn of praise which introduces the 2015

Islamic Declaration on Global Climate Change also opens with an affirmation of our duty to protect and heal 'the bounties of the Lord: Our species, though selected to be a caretaker or steward on the Earth, has been the cause of such corruption and devastation on it that we are in danger of ending life as we know it on our planet.'

This world is but a dew drop world….and yet…. focuses on Buddhist, Taoist and Confucian teachings about impermanence and interdependence; contemplation and engagement; compassion and generosity. The chapter title derives from the writing of poet and ecological activist Gary Synder on the ways in which Buddhist and Taoist insights can help us hold together two apparently contradictory ways of being in the world: to let go of our attachment to the ultimately futile goal of controlling the uncontrollable while continuing to act with courage, compassion and generosity to reduce suffering in the time and place in which we currently exist. It may well be true, Snyder argues 'that it's already far too late to have any effect on the progress of climate change and its effect on ecosystems and human populations…..Yet, still, every day, I feel gratitude to this world that is. [As] Issa's haiku goes: 'This dewdrop world is but a dewdrop world… and yet….' [21, p. 15].

Chapter 9, *Living ecologically* begins by revisiting Rachael Carson's wise counsel on the need to fully understand the delicate interweaving of human and ecological ecosystems. An ever-widening circle of ecologically informed writers including Gregory Bateson, Donella Meadows, Arne Naess, Timothy Morton, Bruno Latour and Val Plumwood continue to build on Carson's foundational ideas about the fragility and complexity of ecological relationships.

The work of these writers is also informed by a strong ethical and practical commitment to develop and articulate social, economic and political paradigms underpinned by heightened awareness of the importance of treating other human beings and other species with justice, compassion and respect. Deepening our awareness of ecological complexity, fragility and beauty can also open our eyes to previously unexplored and unseen landscapes of joyfulness and healing.

Chapter 10, *Action and imagination in the long emergency* is informed by the view that facing a harsh climate future with wisdom and courage depends to a significant extent on our capacity to imagine, design and build the bridges leading to a just and resilient zero-carbon future, and to do so at emergency speed. Visualising the shape and arc of these bridges becomes more difficult every year as emissions continue to rise, triggering increasingly dangerous climate tipping points.

The COVID-19 pandemic has provided a timely wake-up call about the speed with which unexpected shocks can derail the most confident predictions. The most likely outcome of current climate trends remains an increasingly challenging and environmentally hostile world. The likelihood of this outcome intensifies the urgency of identifying social and political tipping points with the greatest potential for accelerating transformational change while renewing and sustaining daily practices of creativity, courage and compassion.

The *2050 Zero-Carbon World Oration* thought experiment provokes us to reflect on the reality that even the most rapid and decisive climate change action is still likely to lead us to a world far more challenging than the planet on which human civilisations have developed and flourished. Looking back from 2050, this story is told through a fictional Oration delivered by Professor Teuila Apatu, Director of the *Global Institute for Climate and Energy Transitions*, reviewing the key events, decisions and actions which drove the great twenty-first-century energy transition at remarkable speed and scale.

In celebrating and honouring the impressive progress made by 2050 towards achieving the global goal of zero net emissions Professor Apatu also notes the severe ecological damage and human suffering caused by the failure to reduce emissions at sufficient speed in the first quarter of the twenty-first century—and the continuing challenge of implementing the actions required to bring global temperatures back below 1.5 degrees. As she notes in her concluding remarks, 'despite the passionate commitment and creativity of so many inspiring scientific, political, business and community leaders over the last 50 years I am still unable to provide my grandchildren the simple, essential gift I most wish to give them: the ecological conditions which enable human beings to continue to thrive and prosper—alongside the many species with which we share this extraordinary planet.'

While it is tempting to leave readers with a neat set of steps ensuring we can indeed pass on the gift of a just and resilient safe climate future, this is a book of reflections and conversations not a road map or manifesto. Some common threads running through these conversations and outlined in the concluding chapter, *The laughter of children, the roar of the ocean* are however clear and strong. Our responsibility to speak with honesty and clarity and power about the climate and ecological emergencies now unfolding around us. The necessity and urgency of accelerating decisive and strategic climate action. The importance of good companions, care and fellowship. The abiding power of justice and compassion; creativity and imagination. And the comfort and delight we still find in celebrating and honouring the beauty of the Earth.

2

Beyond Denial and Despair: Honesty and Action on the Climate Change Front Line

In 2014, science communicator Joe Duggan invited climate scientists to write to him with their responses to the question: How do you feel about climate change risks and threats? [22] Here are three of the letters Joe received.

> Sometimes I have this dream. I'm going for a hike and discover a remote farmhouse on fire. Children are calling for help from the upper windows. So I call the fire brigade. But they don't come, because some mad person keeps telling them that it is a false alarm. The situation is getting more and more desperate, but I can't convince the firemen to get going. I cannot wake up from this nightmare.
>
> Prof. Stefan Rahmstorf, Head of *Earth System Analysis, Potsdam Institute for Climate Impact Research*
>
> These days whenever I am in a beautiful natural environment (a towering forest, a flower filled coastal headland, a pristine beach….) my sense of joy is tinged with deep sadness. At the terrible impact which climate change could wreak on our wonderful planet and at the thought that my grandchildren may never see this but live in a sadly diminished world if they survive at all.

Prof. Ann Sanson, *Department of Paediatrics, Royal Children's Hospital,* Melbourne

How do I feel about climate change? The feelings that seem most immediate, and palpable are those of deep concern, sadness, at times real angst and exasperation at the seeming inability of governments to appreciate and address what is at stake here and the sense that this is a cross-species existential crisis in myriad ways with profound implications for life on Earth as we know it.

Prof. Joe Reser, *Department of Psychology, Griffith University,* Queensland

Responses like these from climate scientists, activists and front-line workers confirm learning from psychological research and our own experience that the first key step in dealing constructively with climate risk and danger is well-informed understanding of the full extent of the threat. In the end, as American novelist James Baldwin rightly notes 'not everything that is faced can be changed. But nothing can be changed until it has been faced' [23, p. 11]. These responses also resonate strongly with the constant struggle many of us face in dealing with the harsh realities of the climate emergency with honesty and courage rather than paralysis and despair.

This brief review of climate trends and tipping points is therefore followed by consideration of the various ways in which denial and avoidance limit our commitment and capacity to take decisive action at sufficient speed. While there is no lack of knowledge about actions required to reduce climate risk, the growing likelihood that global temperature rises will trigger catastrophic ecological tipping points is a source of deep distress for many people working on the climate change front line. There is much we can learn from psychologists and health workers; scientists and activists about strategies for strengthening emotional resilience and for sustaining individual and collective action.

Tell Them About the Water: Climate Change as an Existential Risk

Growing awareness that all evidence is contestable and all claims about evidence and expertise are socially framed does not imply or require an inevitable descent into extreme relativism [24]. This awareness does however strengthen our understanding that the creation and communication of scientific 'facts' is not a simple, linear process of producing and sharing objective 'truths.' Our understanding of climate risk therefore rests on the same foundations as all other learning: information and evidence filtered, sorted and analysed through multiple world views and belief systems.

In an age of competing and contested claims about truth and consequences, many of us find evidence from a variety of sources increasingly valuable in strengthening confidence about the extent and sources of risk and in prioritising action. The scientific evidence about climate change trends and causes, risks and implications is compelling. So too however is the lived experience of Indigenous communities; farmers and firefighters; health and emergency service workers.

For some the analysis and findings from peer-reviewed scientific research and the world's most respected scientific organisations provide a sufficiently convincing case. Here, for example, are some of the key messages from the 2019 *United in Science High Level Synthesis Report* on climate science trends and implications [25]. This report brings together expertise and analysis from many thousands of eminent climate scientists including from *the World Meteorological Organization, the United Nations Environment Program*, the *Intergovernmental Panel on Climate Change*, the *Global Carbon Project* and *Future Earth*.

- Average global temperature for 2015–2019 is on track to be the warmest of any equivalent period on record. It is currently estimated to be 1.1 °C above pre-industrial (1850–1900) times and 0.2 °C warmer than 2011–2015.

- CO2 emissions from fossil fuel use continue to grow by over 1% annually and 2% in 2018 reaching a new high. Despite extraordinary growth in renewable energy, fossil fuels still dominate the global energy system.
- Global emissions are not estimated to peak by 2030, let alone by 2020.
- Observations show that global mean sea-level rise is accelerating and an overall increase of 26% in ocean acidity since the beginning of the industrial era.
- Growing climate impacts increase the risk of crossing critical tipping points. Climate impacts are hitting harder and sooner than climate assessments indicated even a decade ago.

In his introduction to the 2020 update of the *United in Science* climate science synthesis report, United Nations Secretary General Antonio Guterres added these observations.

> This has been an unprecedented year for people and planet. The COVID-19 pandemic has disrupted lives worldwide. At the same time, the heating of our planet and climate disruption has continued apace. Record heat, ice loss, wildfires, floods and droughts continue to worsen, affecting communities, nations and economies around the world. Furthermore, due to the amount of greenhouse gases emitted in the past century, the planet is already locked into future significant heating. [26]

Others of us may find the testimony of individuals and communities with direct experience of climate change-related impacts more compelling. Here, for example, is the call to action which Yeb Sano, lead climate negotiator for the Philippines delivered to the 2013 Warsaw Climate Summit. Following his account of the terrible devastation caused to his country by Typhoon Haiyan Sano challenges climate change deniers to leave their armchairs and ivory towers

> I dare you to go to the islands of the Pacific, the islands of the Caribbean and the islands of the Indian ocean and see the impacts of rising sea levels; to the mountainous regions of the Himalayas and the Andes to see communities confronting glacial floods, to the Arctic where communities grapple with the fast dwindling polar ice caps, to the large deltas of the

Mekong, the Ganges, the Amazon, and the Nile where lives and livelihoods are drowned, to the hills of Central America that confront similar monstrous hurricanes, to the vast savannas of Africa where climate change has likewise become a matter of life and death as food and water becomes scarce.....And if that is not enough, you may want to pay a visit to the Philippines right now..... [27]

Two years later at the *Paris Climate Summit* Marshall Islands poet, Kathy Jetnil-Kijiner, read these words from her poem, *Tell Them* to the assembled delegates.

> Tell them about the water
> how we have seen it rising
> flooding across our cemeteries
> gushing over our sea walls
> and crashing against our homes
> tell them what it's like
> to see the entire ocean level with the land
> tell them
> we are afraid
> tell them we don't know
> of the politics
> or the science
> but tell them we see
> what's in our own backyard. [28, pp. 64–67]

Or, for a view from a more affluent society here are the words which Australian author Richard Flanagan used to describe his response to the heatwaves, fires and floods which swept over Australia in February 2019.

> What has become clear over these last four weeks across this vast, beautiful land of Australia is that a way of life is on the edge of vanishing. Australian summers, once a time of innocent pleasure, now are to be feared, to be anticipated, not with joy but with dread, a time of discomfort, distress and, for some fear that lasts not a day or a night but weeks and months. Power grids collapse, dying rivers vomit huge fish kills while in the North in Townsville there are unprecedented floods and in the south heat so extreme it pushes at the very edge of liveability. [29]

While keeping in mind these caveats about diverse ways of collecting and communicating climate change evidence, a brief overview of scientific findings about climate trends and risks is still a key foundation for understanding the consequences of failing to act with sufficient speed. Careful and critical reading of the vast body of climate science evidence leads most well-informed observers to reach similar conclusions to those of Sir David Attenborough. 'Right now, we are facing a man-made disaster of global scale. Our greatest threat in thousands of years. Climate Change. If we don't take action the collapse of our civilisations and the extinction of much of the natural world is on the horizon' [30].

The 2015 Paris Agreement endorsed the goal of 'limiting global temperature increase to well below 2°C, while pursuing efforts to limit the increase to 1.5°C' [31]. Climate policy ambition and action will however need to rapidly accelerate if there is to be any hope of achieving this target. Climate modelling scenarios in the *2014 IPCC Fifth Assessment* Report indicate a range of likely increases in global surface temperature by 2100 from over 4 °C at the high end (if emissions continue to increase at the rate at which they have over the last 35 years) to 1.5 °C at the low end (if extremely early action is taken to reduce emissions) [4]. *The United in Science Report* noted above concludes that the current level of national climate action commitments and ambition needs to be roughly tripled for emissions reduction to align with the 2 °C goal. They will need to increase fivefold to achieve the 1.5 °C goal [25].

Noting that policies being implemented by many countries are not yet fully consistent with their national pledges, independent climate science research group, *Climate Action Tracker* estimates global warming outcomes from the successful implementation of current (2020) national climate policies in a range from 2.5 to 4.7 °C with a median outcome of 3.4 °C [32]. Even higher levels of risk need to be considered if we take account of the ways in which the full extent of global warming is being masked (by up to 0.9 °C) by the presence in the atmosphere of aerosols such as sulphates, nitrates and dust. As necessary action is taken to reduce these harmful air pollutants, the rate of global warming is likely to further accelerate [33].

It would also be unwise to take too much comfort from the view that many of these global temperature projections are for the end of the twenty-first century. The *Climate Action Tracker* analysis indicates, for example, a high likelihood of global temperatures rising to over 2 °C by 2030 and by over 2.5 °C by 2050 [32]. Likely consequences of global temperature increases of over 1.5 °C (probably already locked in and likely to occur by the early 2030s) include the following [34].

- Destructive and irreversible impacts on Arctic ecosystems (including sea ice decline and thawing of permafrost); mountain ecosystems (including accelerated glacier melting) and warm water coral ecosystems (including coral death and bleaching).
- Destructive and irreversible impacts on human health and livelihoods from extreme weather events such as heat waves, drought, wildfires, heavy rain and flooding as well as from increased water stress and reduced crop production.
- Escalating risks of triggering climate tipping points including the disintegration of the Greenland and West Antarctic ice sheets leading to large and rapid sea-level rise. Sea-level rise of over one metre is likely to create a high level of global risk to all coastal ecosystems and communities.
- The risks to marine ecosystems from ocean acidification (as a result of increased carbon dioxide being emitted into the atmosphere and being absorbed into the oceans) are also rising rapidly with risks becoming 'extremely high' once carbon dioxide concentrations exceed 500 ppm.

Climate Action Tracker's 2020 analysis highlighted the very real possibility that global warming could be as high as 4.7 °C by 2100. When asked to summarise the difference between a world of 2 °C warming and a world in which global warming passes 4 °C, Prof. John Schellnhuber, former Director of the *Potsdam Institute for Climate Impact Research* provides the following blunt assessment. The difference between a 2 °C world and a 4 °C world is, he argues, 'human civilisation' [35]. His response to the question, 'what level of population could the Earth support if global warming exceeded 5°C?' leads to a response which also merits careful reflection: 'below one billion people....' [36]. Schellnhuber goes on to note that a world in which global temperatures exceed 5 °C would in fact mean even higher temperatures in some regions because the land heats up faster than the oceans.

> The continents would be warmer by eight or nine degrees, and highlying regions like the Tibetan Plateau would warm by 12 degrees. All the glaciers would melt. These glaciers feed rivers that sustain two billion people, and they would run dry in the summer. By 2070-80 it would be a problem you couldn't solve anymore....If you do a sober calculation of what the world would look like with unmitigated climate change, you are simply terrified. It's so serious that it's foolish, even infantile to suggest that we will just adapt. [37]

Prof. Kevin Anderson, Chair of Energy and Climate Change at the *University of Manchester* arrives at a similar conclusion. 'A 4°C world is incompatible with an organized global community' [38]. Rachel Warren, coordinator of Land and Water Research at the *UK Tyndall Institute for Climate Change Research* makes this stark assessment. 'In such a 4°C world, the limits for human adaptation are likely to be exceeded in many parts of the world, while the limits for adaptation for natural systems would largely be exceeded throughout the world. Hence, the ecosystem services upon which human livelihoods depend would not be preserved' [39, p. 234].

A planet on which 'the ecosystem services upon which human livelihoods depend would not be preserved.' A world in which global population falls from eight to one billion. An environment 'incompatible with an organised global community' and 'beyond human civilisation.' The real and present danger that current emissions pathways will lead us down the pathway to a future of this kind would surely appear to meet the criteria of 'existential risk.' US President Joe Biden certainly now seems to agree with this view in his 2020 assessment of climate change as 'the existential threat of our time' [40].

An existential risk has been defined, by Professor Nick Bostrom, Director of the *Future of Humanity Institute* at *Oxford University*, as 'one where an adverse outcome would either annihilate Earth-originating intelligent life or permanently and drastically curtail its potential' [41, p. 3]. Bostrom suggests that existential risks are qualitatively different to other kinds of risks. 'Because the consequences are so severe — perhaps the end of human global civilisation as we know it — even for an honest, truth-seeking, and well-intentioned investigator it is difficult to think and act rationally in regard to global catastrophic risks and existential risks' [42, p. 9]. These observations help us understand the puzzling tendency for many of us to seek out a wide array of arguments for avoiding the action required to reduce climate risks capable of drastically curtailing the future of human livelihoods and human civilisation.

Beyond Denial: The Scientific and Ethical Case for Swift and Decisive Climate Action

There is now an extensive literature on the reasons why so many people oppose or try to avoid thinking about the need for urgent and decisive climate action [43–45]. Some sources of climate denial are clearly driven by the desire of vested interests in the fossil fuel, media and finance industries to prevent and delay climate action by infecting public debate with doubt and confusion [46]. The messages that climate change is not 'real' or not caused by human action often also fall on fertile ground among individuals suspicious of evidence produced by 'scientific elites' or concerned that climate action creates unacceptable threats to individual liberties.

A third influential source of climate denial comes from individuals holding deep-seated religious and political beliefs about the right of human beings to exploit and 'subdue' other species. Former Australian Prime Minister Tony Abbott drew heavily, for example on his conservative Catholic values in his 2017 speech to the UK based climate denial think tank, the *Global Warming Policy Foundation*. 'Only societies that have forgotten the scriptures about man being charged with subduing the Earth and all its creatures could have made such a religion out of climate change' [47].

Clear communication of the most robust evidence of climate change trends, causes and risks remains a key foundation for overcoming denial and strengthening understanding of the urgency of action. Evidence of the ways in which climate trends are increasing the frequency and severity of extreme weather events is particularly important in enabling us to join the dots between personal experience and broader climate trends.

Our interpretation of the implications of climate science evidence and messaging is however also heavily influenced by pre-existing value frameworks and political perspectives. For some of us, an ethical concern about the consequences of catastrophic climate risks for the most vulnerable people and species is sufficient motivation. For others, recognition of more immediate and personal risks to our own families and communities will be crucial. So too will heightened awareness of the social and economic opportunities and co-benefits of a just and sustainable post-carbon future. The views of many government, defence and business decision makers are also increasingly influenced by their assessment of the grave security, financial and economic risks of climate change as well as the huge economic opportunities opening up in shifting away from a fossil fuel-based economy.

Outright rejection of climate science is becoming harder to sustain in the face of personal experience of the frequency and severity of extreme weather events. Denial of the reality of anthropogenic climate change is however being replaced by more insidious and seductive denial of the speed and scale of action required to adequately address climate risk. 'Alex Steffan, author of *World Changing: A Users Guide for the 21st Century* describes this new form of climate denial as 'climate gradualism,' the idea that 'we all want to act on climate, but we also have to be slow, incremental and realistic' [48].

As Australian climate activist and author, Paul Gilding also notes, the dangerous consequences of glossing over existential risks have been demonstrated yet again in the slow and clumsy response by many governments to COVID-19. We keep assuming, Gilding suggests 'these events are somewhere "out in the future" and "we always figure these things out". The pandemic was like that. Until it wasn't. And we didn't. Observe the consequences' [49].

Common arguments for and against delaying decisive climate action include the following.

Emissions reductions at the required speed and scale would have too great an impact on economic growth, prosperity and wellbeing—particularly for the poorest and most vulnerable communities. The argument that rapid emissions reductions are likely to have too great an impact on developing countries sounds a little less self-serving when this view is expressed by citizens from those vulnerable communities rather than senior executives and wealthy shareholders of global mining companies. The CEOs of many companies are however increasingly aware that the investment opportunities arising from a rapid transition to a zero-carbon economy far outweigh the costs.

The most threatening climate risks are still some way off in the future or happening far away. There are many more immediate competing priorities: earning a living, keeping the crops alive; looking after children and aging parents; replacing the roof after the latest severe storm. The problem here, George Marshall, author *of Don't Even Think About It: Why Our Brains Are Wired to Ignore Climate Change* notes is the way in which the 'creeping problem' of climate change provides many of us with a comforting excuse for postponing action. 'We allow just enough history to make it seem familiar but not enough to create a responsibility for our past emissions. We make it just current enough to accept that we need to do something about it but put it just too far in the future to require immediate action' [50, p. 64].

The climate challenge is indeed daunting but the human capacity for ingenuity and technological innovation will ultimately create the necessary solutions. Look at the remarkable advances in renewable energy and energy efficiency; electric vehicles and smart cities. Even in the worst-case scenarios a range of geoengineering strategies can be deployed to directly cool the planet and to capture and store CO_2. Faith in the power of technology to overcome every human challenge is a tempting and seductive narrative. Some of us wonder however as we reflect on the dystopian ecological consequences of unconstrained economic growth whether we would be wise to remember the price which Prometheus paid for his hubris in stealing fire from the Gods. Others of us remain doubtful about the wisdom of geoengineering strategies such as pumping sulphur dioxide into the atmosphere without considerably greater understanding of the consequences and implications of such action.

Climate change is primarily the result of capitalist institutions and relationships and of actions by the wealthy and powerful. An overemphasis on impending climatic and ecological catastrophes diverts attention from the more immediate political priority of ending unequal and oppressive power relations [51]. While evidence of the link between capitalist power relations and the underlying drivers of global warming is compelling the problem of urgency remains. Is it plausible to imagine a fundamental shift in capitalist power relations in a time frame consistent with the speed and scale of emissions reductions required to prevent catastrophic climate change?

It is now too late to prevent catastrophic climate change. We should therefore enjoy life while we can and focus on protecting our own families, friends, communities and nations. At an individual level, this argument is consistent with various forms of 'seize the day' hedonism and of survivalist strategies preparing for dystopian futures. In national security and military circles, it can also quickly become an additional justification for strengthening investment in the military preparedness necessary to protect national borders from an influx of potentially desperate 'climate refugees.'

While noting the strong hold these arguments still have on many of us it is also important to recognise the growing breadth and depth of concern about climate change threats and consequences. In 2020, the annual *Pew Research Centre* survey on global attitudes found that 70% of people identified climate change as 'a major threat' to their country, up from 56% in 2013 [52]. In Australia, the 2019 *Lowy Institute* survey found that 64% of Australians saw climate change as 'a critical threat,' up from 48% in 2014. Over 60% of Australians held the view that 'global warming is a serious and pressing problem and we should begin taking steps now even if this involves significant costs, an increase of 25 percentage points since 2012' [53].

Pathways to a Just and Resilient Zero-Carbon Future

There is now widespread understanding that keeping global temperatures close to 1.5 °C will require greenhouse gas emissions to peak by the early 2020s followed by rapid reduction to as close to zero as possible by 2040 [54]. The actions which need to be taken to achieve these goals are also clear. Replacing fossil fuels with renewable energy. Reducing energy demand by reducing consumption and improving energy efficiency. Reducing emissions from agriculture, forestry and land clearing; and drawing down CO_2 emissions in ways consistent with social justice and the health and wellbeing of all communities and ecosystems [2]. Massive investment in climate resilience infrastructure and capability will also be an essential foundation for meeting the challenges of climate impacts already locked in.

Technological innovation and falling costs do indeed continue to drive remarkable growth in renewable energy. Global renewable energy capacity grew by over 8% per year between 2011 and 2018 [55]. Between 2010 and 2019, the cost of solar PV fell by over 80% and the cost of onshore wind by over 40% [56]. In 2018, the *International Energy Agency* predicted that renewables would continue to grow strongly providing 30% of global power demand by 2023 [57].

The problem of course is that, while the share of energy provided by renewables continues to accelerate, fossil fuels still provide well over 80% of global energy and CO_2 emissions continue to rise. While reducing energy demand and improving energy efficiency have huge potential to accelerate the shift to a low carbon economy, energy efficiency improvements still risk being overwhelmed by rising energy demand, particularly from emerging economies. Recent trends in agriculture and forestry and land use are following a similar path with gains in low carbon agriculture offset by increasing deforestation [58].

The roadblocks preventing rapid and effective emissions reduction action are all well-known: the power and influence of the fossil fuel industry and other vested interests; social and technological path dependencies; financial, governance and implementation constraints; the dominant neoliberal economic paradigm of unsustainable consumption and inequitable wealth distribution; and the paralysing consequences of despair. Removal of these roadblocks is clearly an essential foundation for accelerating climate action.

Strategies for reducing the influence of coal, oil and gas corporations include strengthening awareness of the risks which businesses, governments and communities are exposed to in failing to address climate change and transition impacts; ending public subsidies for fossil fuel industries; and decisive, enforceable regulatory action driving investment from fossil fuels to renewables. Enabling and supporting an equitable, just transition for communities and households affected by the phase out of fossil fuels will also be vital, both for ethical reasons and to strengthen public support for the implementation of tough political decisions.

Shared narratives and examples of successful local and regional energy transition strategies can help communities visualise the desirability and feasibility of alternative pathways to a just and resilient zero-carbon future. Strong government leadership in setting science-based emission reduction targets, ramping up carbon price levels and mobilising the investment required to scale up commercialisation and deployment of game-changing social and technological innovations will all be important tools for overcoming path dependencies and lock ins.

Overcoming governance and implementation constraints will require sustained reinvestment in public sector capabilities and skills; collaborative alliances and information sharing between nations, regions and cities and an increased role for local government and community organisations in creating innovative transition solutions. Public support for accelerating climate action can be further strengthened by expanding opportunities for citizens to participate in informed debate about energy transition risks and opportunities.

Many authors have pointed to the unsustainability of dominant economic paradigms based on the assumption of unconstrained consumption of resources and energy on a finite planet [59]. A substantial reduction in the consumption of energy and resources will clearly be an essential precondition for the rapid transition to a zero-emissions global economy. Achievement of this aim will also require ongoing and concerted action to increase confidence that this transition can be achieved in ways consistent with the goals of social, economic and climate justice.

Beyond Despair: Wisdom and Resilience in the Long Emergency

The first and most obvious psychosocial consequences of climate crisis are the direct impact on mental health and wellbeing of increasingly frequent and intense extreme weather events and disasters such as storms, floods, fires and droughts. Trauma and stress from such events are likely to be greatest for vulnerable and low-income individuals and communities including children and older people; racial and ethnic minorities and people with pre-existing mental health issues [60]. Many vulnerable individuals and communities are also experiencing increased trauma, stress and grief as a result of the longer-term implications of climate change for accelerating and intensifying drought and famine; insecurity and violence; military conflict and forced migration.

Heightened understanding of climate risks and tipping points is often experienced most intensely by climate science researchers, activists and emergency service workers on the climate change front line [61]. The potentially paralysing impact of climate grief leads climate activist and author Bill McKibben to the cautionary observation that despair about the difficulty of driving climate action at sufficient speed can easily become a self-fulfilling prophecy. 'My only real fear' McKibben reflects 'is that the reality now increasingly evident in the world around us will be for some an excuse to give up. We need just the opposite – increased engagement' [62].

Recent research by social scientists Lesley Head and Theresa Harada explores the diverse array of coping strategies employed by climate researchers and policy makers to overcome despair and sustain emotional resilience [63]. Regular exercise, yoga and meditation; art and music and the healing power of the natural world all play important roles. Some climate scientists also employ deliberate strategies of emotional distancing, focusing tightly on the quality and rigour of their scientific work and maintaining a strict separation between their research and other aspects of personal and family life. One of the scientists interviewed described their strategy for separating work and non-work life in the following way. 'I actually deliberately block it out as much as I can. I don't think I'd be in very good shape if I let myself think about it all the time. So there's a denial for you [laughing] I have to say to be brutally honest I tend not to look at the future too much because it's very confronting. I have kids' [63, p. 97].

Black humour also forms part of the repertoire some scientists use to sustain action and to deflect the impact of continuing attacks from climate denialist trolls. 'You can't have kept on doing this for a long time without having a bit of graveyard humour. If you let it get you down then you're gone, you're no longer with us. The ones that have remained are the ones who are who are sufficiently stoic about this to laugh off the regular attacks' [63, p. 97].

Stoic strategies emphasising the wisdom of only worrying about the sources of problems we can in fact change are common elements in many psychological guides for coping with eco-anxiety and climate grief. More nuanced stoic insights about the power of the mind to reframe the stories we tell ourselves have also played a foundational role in the development of the dominant therapeutic paradigm of our time, Cognitive Behavioural Therapy (CBT) [64].

CBT informed strategies and techniques inform and underpin the advice provided by many psychologists and therapists to people experiencing climate change-related distress and anxiety. These psychological coping strategies also commonly include balancing our commitment to urgent and decisive action with a sensible awareness of the capacity of any individual to solve a problem as vast and complex as climate change; staying well-informed while avoiding being overwhelmed by social media and the 24/7 news cycle; sharing our grief and anxiety with trusted friends and family members; maintaining healthy routines of good food, good sleep, meditation and exercise; valuing and practising kindness and compassion for ourselves and each other; and spending time in calm and restorative environments.

These are clearly all sensible and useful strategies for dealing constructively with climate grief and eco-anxiety. Psychologists Marc Pilisuk and Jamie Rowen draw on the experience of nuclear disarmament activists to suggest that holding our nerve in the face of an approaching firestorm is also easier when we stand together, shoulder to shoulder with others committed to taking decisive and effective collective action. 'Participation in activity with others can make people more powerful in matters that count in their roles as parents and as citizens and give them a psychological sense of community and of empowerment. Large threats to humanity can only be addressed by cooperative action and by working together with other caring people. This can be as rewarding as it is empowering' [65, p. 7]. The contribution which work with others can make in strengthening emotional and psychological resilience as well as in driving transformational change is therefore the main focus of the next chapter: *Remembering Magnificence: Collective Action and the Beauty of the Earth.*

3

Remembering Magnificence: Collective Action and the Beauty of the Earth

Walking with good companions is, as historian Howard Zinn recalls, often the best place to start a journey in dark times.

> To be hopeful in bad times is not just foolishly romantic. It is based on the fact that human history is a history not only of cruelty, but also of compassion, sacrifice, courage, kindness. If we remember those times and places – and there are so many – where people have behaved magnificently, this gives us the energy to act, and at least the possibility of sending this spinning top of a world in a different direction. [66]

Human history is full of extraordinary stories of courageous collective action driving transformational change which few at the time saw coming: the abolition of slavery, the triumph of the Suffragettes, the overthrow of apartheid, the fall of the Berlin Wall, the independence of Timor Leste. Reflection on stories such as these leads historian Rebecca Solnit to the following observation about the value of remembering and honouring courageous action.

Together we are very powerful, and we have a seldom-told, seldom-remembered history of victories and transformations that can give us confidence that, yes, we can change the world because we have many times before. You row forward looking back, and telling this history is part of helping people navigate toward the future. We need a litany, a rosary, a sutra, a mantra, a war chant of our victories. The past is set in daylight, and it can become a torch we can carry into the night that is the future. [11]

The first and most urgent priority in facing the existential threat of climate emergency is to further accelerate actions being taken all around the world to drive the transition to a just and resilient zero-carbon economy. Awareness of the passion, creativity and energy of the millions of individuals working together to achieve this transformation is also a potent antidote to paralysis and despair.

The climate action initiatives outlined here provide seven short glimpses of the vast body of inspiring work being undertaken by millions of people in many different cultural and political contexts. While the actions described here demonstrate varying levels of impact and success they each in different ways illustrate the potential for participation in collective action to overcome isolation and despair by strengthening a sense of shared concern and purpose. They also demonstrate the remarkable speed with which well-focused and well-timed collective action can trigger swift and transformational social and political change.

Contemplating the Beauty of the Earth: Rachel Carson and *Silent Spring*

The first book I took down from the shelf when I began this project was Rachel Carson's *Silent Spring* [67]. First published in 1962, *Silent Spring* is widely known as the wake-up call we needed to warn us of the risk which unconstrained use of pesticides like DDT posed for birds and insects. Many also regard the publication of *Silent Spring* as a foundational moment in the birth of the global environment movement.

Rachael Carson's life and work exemplifies and illustrates many of the qualities of 'magnificence'—of 'compassion, sacrifice, courage and kindness' referred to by Howard Zinn. Her scientific rigour and skilful advocacy, often in the face of brutal opposition continue to inform the core values and strategies of modern ecological and climate action movements. The theoretical and scientific implications of Carson's work for deepening our understanding of the interdependence of human beings and nature are explored more fully in Chapter 10. The main focus here is on her contribution to our awareness of the role which delight and joy in the wonders of the Earth can play in inspiring and sustaining individual and collective action.

Carson's work as a biologist, science communicator and activist was constantly inspired and energised by her deep enjoyment and fascination with the mysteries and splendour of ocean life. Here, for example, are the opening lines of her article *Undersea*, initially written as the introduction to a 1935 *US Bureau of Fisheries* brochure and first published in *The Atlantic* in 1937.

> Who knows the ocean? Neither you nor I, with our earth-bound senses, know the foam and surge of the tide that beats over the crab hiding under the seaweed of his tide pool home; or the lilt of the long, slow swells of mid-ocean, where shoals of wandering fish prey and are preyed upon, and the dolphin breaks the waves to breathe the upper atmosphere. Nor can we know the vicissitudes of life on the ocean floor, where the sunlight, filtering through a hundred feet of water, makes but a fleeting bluish twilight, in which dwell sponge and mollusk and starfish and coral, where swarms of diminutive fish twinkle through the dusk like a silver rain of meteors. [68, p. 4]

In rereading these words, I remember vividly my own childhood encounters with the astonishing beauty of the oceans and rivers; mountains and forests of the small corner of southern Australia in which I have had the good fortune to be born. Like many other climate scientists and activists, my gratitude for these gifts is troubled and tempered by awareness of the speed with which these fragile worlds are being overwhelmed by heatwaves, fires and storms. Carson's decision to introduce the carefully evidenced alarms and warnings of *Silent Spring* in the following way was no doubt informed by similarly mixed emotions.

> There was once a town in the heart of America where all life seemed to live in harmony with its surroundings. The town lay in the midst of a checkerboard of prosperous farms, with fields of grain and hillsides of orchards, where, in spring, white clouds of bloom drifted above the green fields. In autumn, oak and maple and birch set up a blaze of colour that flamed and flickered across a backdrop of pines. Then foxes barked in the hills and deer silently crossed the fields, half hidden in the mists of the fall mornings. Then a strange blight crept over the area and everything began to change. [67, p. 1]

Recent scientific research about the emotional and psychological benefits of the time we spend in 'green spaces' continues to confirm the accuracy and wisdom of Rachel Carson's reflection that 'those who contemplate the beauty of the Earth find reserves of strength that will endure as long as life lasts. There is symbolic as well as actual beauty in the migration of the birds, the ebb and flow of the tides, the folded bud ready for the spring. There is something infinitely healing in the repeated refrains of nature — the assurance that dawn comes after night, and spring after the winter' [69, p. 89].

Sharing this quotation with climate scientist and activist colleagues sometimes leads to a wry smile and the observation that the assurance that spring will follow winter is becoming highly questionable in a world of increasingly unpredictable, increasingly daunting climate trends and tipping points. Carson, I suspect might nod her head in gentle agreement, offering by way of comfort the following concluding reflections about interdependence and ephemerality from her 1937 article *Undersea*.

> Thus we see the parts of the plan fall into place: the water receiving from Earth and air the simple materials, storing them up until the gathering energy of the spring sun wakens the sleeping plants to a burst of dynamic activity, hungry swarms of planktonic animals growing and multiplying upon the abundant plants, and themselves falling prey to the shoals of

fish; all, in the end, to be redissolved into their component substances when the inexorable laws of the sea demand it.

Individual elements are lost to view, only to reappear again and again in different incarnations in a kind of material immortality. Kindred forces to those which, in some period inconceivably remote, gave birth to that primeval bit of protoplasm tossing on the ancient seas continue their mighty and incomprehensible work. Against this cosmic background the life span of a particular plant or animal appears, not as a drama complete in itself, but only as a brief interlude in a panorama of endless change. [68, p. 12]

The inevitably of endless change does not, Carson would no doubt hasten to add, mean that we should step back from our determination to reduce the catastrophic impacts of unconstrained consumption and exploitation of the Earth's resources. Awareness of our fragility and impermanence does however strengthen understanding of the need to create sustainable foundations for the individual and collective action required to address overlapping and compounding emergencies which it is now clear will continue to unfold over many centuries.

The Only Limitations We Have Are Our Own Imaginations: *350.org*

In 2005, a small group of students at Middlebury College, Vermont began a conversation with one of their lecturers, environmental writer and activist Bill McKibben about strategies for building growing concern about climate change into a social movement of real size and power [70]. One of the students, May Boeve recalls this conversation as reinforcing her sense that 'a lot of people were concerned about climate change. But it didn't look like the movements we'd studied in school, with protests and songs and visual imagery and analyses of power' [71]. In the autumn of 2006 on returning from one of their first collective actions, a five-day climate change protest march across Vermont, McKibben and the students he walked with were surprised to find the media describing their march of a thousand people as the largest ever climate change demonstration in the United States. If this was indeed the largest climate demonstration ever, there was clearly much work to be done.

By the spring of 2007, the newly formed *Step It Up* campaign had coordinated 1400 climate protests across all fifty states of the United States. In 2008 the group changed its name to *350.org*, a number widely understood as the safe level of atmospheric CO_2 parts per million. In October 2009, the first *350.org International Day of Action* was organised to inspire delegates attending the COP 15 climate conference in Copenhagen to focus on the goal of an 80% reduction in emissions by 2050. The 5200 actions across 181 countries ('the most widespread day of action in the planet's history' according to CNN) were indeed unprecedented and inspiring in their breadth and creativity.

May Boeve recalls the revitalising energy of working together to turn many thousands of conversations, ideas and plans into highly visible and influential actions all around the world. 'I can't underscore enough how it felt. The late-night meetings and actions and strategy sessions with my friends — it was incredibly exciting' [71]. Fellow *350.org* organiser Jamie Henn added, 'it was all very homemade. It was a learning experience, a community process. There was no grand plan, no level of expertise. We were just naive/ambitious enough to try to do something at a large scale. It took off. People were looking for something' [72].

While *350.org* continued to grow and flourish, with over one million members responsible for many thousands of actions, the failure of COP 15 to drive decisive climate action strengthened the determination of the *350.org* team to further sharpen their goals and strategies. In 2011, *350.org* decided to add their strength to the campaign to stop the Keystone XL Pipeline, designed to transport oil from the Alberta tar sands to oil refineries in Illinois and Texas. Opposition to the Pipeline had been building since 2008, driven initially by concerns from First Nation peoples and farmers about damage to freshwater aquifers. Growing evidence about the importance of keeping oil and other fossil fuels in the ground in order to stay within the global carbon budget was the catalyst for *350.org* to prioritise the campaign to convince President Obama to stop the Pipeline from proceeding.

In reflecting on the request that Keystone XL protestors always wear their 'Sunday best' organiser Jamie Henn noted that, 'we wanted to convey that same level of moral seriousness that the civil rights movement did. That was intentional both because we thought it was important for the situation at hand, and for the power that came from the historical echo' [71]. Four years of Keystone campaigning culminated, on November 6, 2015, in President Obama announcing formal rejection of the Pipeline. The battle continued to rage with President Trump's 2019 decision to approve and accelerate the opening of Keystone followed in 2020 by the Supreme Court decision to continue to block construction of the Pipeline.

In 2012, *350.org* launched the *Go Fossil Free: Divest from Fossil Fuels!* campaign calling for universities, cities, religious institutions and pension funds to withdraw their investments from fossil fuel companies. The campaign logic was sharp and simple. 'If it is wrong to wreck the climate, then it is wrong to profit from that wreckage' [73]. One year later, in 2013, fossil fuel divestment campaigns had been triggered at 308 US colleges and universities, 105 cities and states and 6 religious institutions. By December 2018, the global divestment campaign had led to divestment commitments of $7.93 trillion by over 1000 businesses, university and religious organisations as well as by cities like New York and national governments such as Ireland [74].

Bill McKibben assesses the significance of the divestment movement in the following way. 'Divestment by itself is not going to win the climate fight. But by weakening – reputationally and financially – those players that are determined to stick to business as usual, its one crucial part of a broader strategy.....With activists marching and going to jail, phrases such as stranded "assets" were soon appearing in the mouths of everyone from hedge fund managers to the governor of the Bank of England' [75].

On September 2014, the People's Climate March co-organised by 350.org brought 400,000 people to the streets of New York City. On that day, there were 2500 other climate action demonstrations and events in over 160 countries. In reflecting on her experience of the journey from hundreds of people walking through the forests of Vermont to hundreds of thousands of people marching through the streets of New York, May Boeve noted her concern that the enormous growth in climate activism was not yet matched by policies at anything close to the required speed.

Her hope and courage and energy were sustained however by her sense that there was now much greater clarity about the primary target.

> Now there's widespread agreement about combating the fossil fuel industry all over the world. There's not a whole lot of difference of opinion about whether to do that. There's a whole lot of difference of opinion about strategy, but with that all understood, all we have to do is figure out how to make that happen on the best timeline we can. Not in an arrogant 'we figured it out' way but 'the only limitations we have are our own imaginations.' And what we're capable of.... [71]

We Are Not Drowning, We Are Fighting! the Pacific Climate Warriors

On 17 October 2014, at the Port of Newcastle (Australia's largest coal port) a young Pacific Island woman led thirty of her fellow *Pacific Island Warriors* in prayer as they stood beside their richly decorated, hand-carved canoes. The *Warriors*, supported by hundreds of Australian climate campaigners then paddled their canoes out into the harbour to confront and block coal ships entering and leaving the Port. They also occupied the offices of a number of the banks facilitating the expansion of Australian coal, including the global headquarters of ANZ, a major fossil fuel investor and the primary banking service for many Pacific Islanders. Marshallese climate warrior and daughter of the President of the Marshall Islands, Milan Loeak explained the reason for their actions in this way.

> We're blockading the world's largest coal port in Newcastle to show that elsewhere in the world, whole nations are paying the price for Australia's coal and gas wealth. We're calling on the Australian public to come and join us in saying no to the building of new coal and gas infrastructure. We're imploring you to stand with us, canoe against coal ship, and help us defend our homes against those who would seek to destroy it. Climate change is an issue which will eventually affect every individual on this planet and someday, you may need someone else's help to defend your home too. [76]

The *Pacific Climate Warriors* were formed in 2011 by young people from 15 Pacific Island nations to communicate more widely the vulnerabilities of their island communities while also showcasing their strength and resilience. Tongan *Climate Warrior* Silivesiteli Loloa notes that the *Warriors* reject common depictions of Pacific Islanders as doomed and passive victims condemned to a bleak, inevitable future as climate refugees.

> Tonga is my island paradise. As a Warrior from the kingdom, I have been taught to respect my culture, our traditions, my family and our land. I will do everything that I can to fight for my home and will continue to declare, 'We are not drowning, we are fighting.' There is a world of difference between viewing Islanders as climate-change victims in a far and rising sea and viewing them as a sea of warriors with the power to rise up against climate change. The first emphasizes helplessness and victimization, while the second acknowledges Islanders' agency and ability to work together. [77]

Fijian Climate Warrior and Pacific Communications Coordinator for *350.org*, Fenton Lutunatabua defines the character of a warrior in the following way.

> A warrior is resilient....not aggressive or violent....Assertive....serves to protect their community, culture, land and ocean....Always learning, responds to the needs of those around them and of the greater good non-violently....Stands their ground against an adversary, against injustice and against oppression....Respectfully embodies their local culture and traditions....Is accountable for their actions and words....Serves those who cannot fight for themselves: future generations, animals and plants, environments.... [78]

In 2017, prior to COP 23 in Bonn, the *Pacific Climate Warriors* travelled to Germany to work with communities and activists protesting against the continuation of coal mining in the German Rhineland. Ahead of the protest, the *Warriors* held a traditional ceremony in the nearly deserted village of Manheim, soon to be cleared to make way for an expanded coal mine. Women from the local community took part in the ceremony, receiving a gift of red flower petals on tapa cloth representing the beauty and resilience of Pacific culture. In presenting the gift, 19-year-old Samoan *Climate Warrior* Brianna Fruean explained 'we are here today to jointly demand the end of fossil fuels. In our countries right now climate change is a reality, we live in the eye of the storm, you can't go a day without seeing the decline in fish or the sea level rising or the intense change in weather' [79].

In *Moana: The Rising of the Sea*, a 2014 film made by the *Pacific Climate Warriors*, young Fijian men and women are called from their daily activities by a conch shell warning to go to the shore and prepare for battle [80]. Dressed in richly decorated ceremonial clothes they take up their warrior stance as the waves crash around their legs. The leader greets them with the following call to action. 'Listen to me! Climate has changed. You and me. Let's unite! Are you ready?' The warriors respond with passion and defiance. 'Yes! We are ready!' They then perform a series of dances illustrating the escalating threat of climate change; affirming the warriors as the children of Fiji and of the ocean; joining together in the building of their ocean-going canoes and summoning other Pacific Islanders to join them in collective defiance.

Fenton Lutunatabua's observations on the diverse skills and respectful cooperation required to build and sail an ocean-going canoe provide a striking metaphor for the contributions and approaches required to meet the global challenge of climate change. Building a canoe, he explains usually involved the entire community.

> While there was segregation of roles (men made the hull and rigs, women made the sails), the entire community — men, women, young people and elders — came together to make the rope and cord that bound the craft together. Without the participation of the entire community, the canoe would not be complete: the absence of any part could render the vessel unusable, and the absence of skilled and knowledgeable sailors could make a successful voyage unlikely. On a global level, we must recognize that it will take all of our diverse skills, knowledges, experiences, and resources to put an end to the threat that plagues the Pacific and the world. [81]

We Did This Together. We Did This With Art. We Did This As Art *Liberate Tate*

On the 28th of November 2015, the first day of the COP 21 Paris Climate conference, a new, unauthorised performance exhibition, *Birthmark* opened at London's flagship modern art gallery, *Tate Britain*. The central actor in *Birthmark* was a tattooist who publicly tattooed her fellow artists with a number specifying the amount of carbon dioxide in the atmosphere, expressed as parts per million (ppm), in the year of their birth. A performer born in 1962 was, for example, tattooed with the number 318 (ppm) while a performer born in the year 1993 received the number 357 (ppm).

The *Birthmark* exhibition guide explained that the exhibition was organised by the art collective *Liberate Tate* to communicate the damage being caused by climate change and to continue to pressure *Tate Britain* to end sponsorship from BP and other fossil fuel companies [82].

> **Birthmark:** A live unsanctioned performance by *Liberate Tate* November 2015, Tate Britain. Skin, needles, ink, rotary machine, latex, plastic, bodies, trauma
>
> Climate change is permanent; so are tattoos. This piece explores lasting damage, scarring, and healing. Numbers are written on the body, brands are written on the gallery and, as carbon is released into the atmosphere, damage is written on the planet. In response to climate change, this performance embodies the revisions being inscribed on our planet in an intimate, personal way. Each tattoo echoes the engraving act by the oil sponsor in transforming the body of the gallery. In the run up to the international climate talks in Paris, the artists invite Tate to reconsider their sponsorship deal with BP, and to begin to erase this scar from their skin.

Liberate Tate was founded following a 2010 art and activism workshop which coincided with the crisis of the BP Deepwater Horizon oil rig spilling hundreds of millions of gallons of crude oil into the Gulf of Mexico [83]. The workshop had been officially commissioned by *Tate Britain* and participants were therefore somewhat surprised when *Tate* curators attempted to censor them from making interventions against *Tate* sponsors, even though no such interventions had in fact been planned.

Workshop participants decided, in response, to set up a new art and advocacy collective, *Liberate Tate* to organise acts of 'creative disobedience' in order to end sponsorship of the *Tate* by BP and other fossil fuel companies. Members of the *Liberate Tate* collective noted that many fossil-fuel companies, including BP frequently use art and cultural sponsorship to greenwash their reputations and protect their social license. 'By using the words BP and Art together, the destructive and obsolete nature of the fossil-fuel industry is masked, and crimes against the future are given a slick and stainless sheen' [84].

The *Liberate Tate* activists were also keen to raise broader political and social questions about the relation between fossil-fuel corporations, cultural institutions and climate justice. 'It is equally a concern if ecological issues are seen as divorced from social ones. The very 'specificity' of ecology implies that environmental damage is the only concern, which lets the oil companies off the hook for numerous human rights violations' [85, p. 627].

'Creative disobedience' performances organised by Liberate Tate included *License to Spill*, a calculated 'oil spill' at the 2010 *Tate* Summer Party; *The Gift*, a public presentation of a 16.5-metre wind turbine blade to be donated to Tate's permanent collection; and *Time Piece*, a 25-hour 'textual intervention' that saw the floor of *Tate Modern's* Turbine Hall covered in commentary drawn in charcoal on the relationship between art, activism, climate change and the oil industry.

Liberate Tate was also extremely effective in triggering action by other artists and cultural workers. The *Reclaim Shakespeare Company*, for example, deployed classically trained actors who scripted their protests in iambic pentameter and acted them out on stage in full Shakespearian costumes. Here's a sample of their work.

> The *Royal Shakespeare Company, British Museum, Royal Opera House, National Portrait Gallery*, and *Science Museum* have all chosen to put BP's money in their purse. Yet he's mad that trusts in the tameness of a wolf. BP is doing everything in its power to let not the public see its deep and

dark desires–fossil fuel expansion and ecological devastation. Enough! No more!
Times are tough. Ay, there's the rub. But all that glisters is not gold. And whilst comparisons are odorous, we do well remember the dropping of tobacco companies as sponsors by a host of cultural institutions. The arts continued, and so shall the *British Museum* and the *RSC*, freed from the grasp of this smiling damned villain. Once more unto the breach, dear friends, once more! We believe that action is eloquence. We say to our cultural institutions: to thine own self be true. Be nothing if not critical and forgo your damaging relationship with BP. [86]

The actions of *Liberate Tate* and other allied organisations such as *Art Not Oil, BP or Not BP* and *The Reclaim Shakespeare Company* clearly had an impact. In March 2016, the *Tate* announced it would end sponsorship from BP. *Liberate Tate*'s statement celebrating this announcement provides an eloquent description of the disruptive and creative potential of audacious and well-targeted collective action.

This is a day many thought would never come. BP sponsorship of Tate is inevitable, they said. BP sponsorship of Tate is vital, they said. Without it, Tate could not function, they said. That's just the way it is, they said. But we changed that. You changed that. We all did this. We stayed true to a collective, collaborative artistic practice to create porous, participatory performances that were genuinely confrontational. We did this with our determination, commitment, stamina, tenacity, audacity, outrage, creativity, artistic craft, deep ecology and soulful collaboration……We did this together. We did this with Art. We did this as Art. [87]

By late 2018, a number of other British cultural institutions including the *Edinburgh International Festival*, the *National Gallery*, the *Southbank Centre* and the *Natural History Museum* had ended fossil fuel sponsorship. Successful campaigns against fossil-fuel sponsorship were also conducted against the *Canadian Museum of History, American Museum of History* and the *Van Gogh Museum* in Amsterdam. *Liberate Tate* activist, Mel Evans summarised the impact on the artists of organising these civil disobedience performances and artworks in the following way. 'Its about people and about community building and opening up a space where artists and activists can come together and learn from each other. I think Liberate Tate's work has really been a place where artists who've never taken part in protest or direct action have tried it out – it's been quite a transformational process for artists and for activists' [88].

We Need Social Tipping Points: *The German Energiewende*

The *350.org*, *Pacific Climate Warriors* and *Liberate Tate* initiatives provide three diverse examples of ways in which imaginative and courageous collective action can challenge the power of vested interests and broaden public support for emergency speed climate action. The *German Energiewende* (Energy Transition) illustrates the potential for sustained and concerted action by citizens and governments to move beyond advocacy and begin the long and complex task of decarbonising an entire national energy system [89].

While the jury is still out on the long term outcomes of the *Energiewende* the vast array of energy transition projects explored and championed by German households, communities and governments have already provided a hugely influential 'living laboratory' for communities and governments around the world seeking to imagine and create alternative energy futures [90]. This capacity to create and enact alternative social, political and economic imaginaries is why, as Bill McKibben notes it is possible to argue 'that the Energiewende is the most important new public policy initiative anywhere on the planet in the last fifty years' [91].

The first steps on the *Energiewende* pathway were taken in the 1970s as German anti-nuclear activists began to envisage energy systems capable of providing viable alternatives to nuclear power. Eva Quistorp, a leading figure in the 1970s anti-nuclear movement argues that the involvement of local people from diverse social backgrounds played a key role in broadening support for a decentralised and democratised approach to the German energy transition. 'At the heart of the movement were the farmers, vintners, families, housewives, and parish pastors. Students and experts contributed too, but the movement's force came from self-organized, citizens' initiatives' [89].

In 1980, researchers from the *German Institute for Applied Ecology* brought together key threads from the multitude of energy transition debates and experiments in a book with the prescient and provocative title: *EnergieWende: Growth and Prosperity Without Oil and Uranium* [92]. German public support for alternatives to nuclear power continued to build following the 1987 Chernobyl nuclear disaster. Pressure to accelerate the energy transition was further increased by the 1997 *Kyoto Protocol Agreement* which required Germany to cut CO_2 emissions by 21% by 2012 relative to 1990 levels.

In 2000, the SDP (Social Democrat) and Green Party Coalition Government passed the *German Renewable Energy Act* guaranteeing grid priority and fixed feed in tariffs for 20 years for anyone generating renewable power. This feed in tariff guarantee provided strong incentives for an exponential expansion of household and community investment in solar and wind energy. SDP Member of Parliament Nina Scheer commented that 'no one expected renewables to shoot up so high, so fast. The Act sparked a real grassroots citizen's movement. Germans turned the *Energiewende* into their own project' [89].

In 2010, the Conservative government of Angela Merkel released the *German Energy Concept* further strengthening renewable energy and energy efficiency targets and investments [93]. While the *Energy Concept* included initial support for nuclear power, this position was quickly reversed following the melt down of the Fukushima power station. From 2011, Chancellor Merkel consistently referred to the *Energiewende* in describing Germany's energy transition goals and projects. Despite the ongoing challenge of maintaining precarious Conservative and Social Democrat coalitions the 2016 *Climate Action Plan 2050* continued to ramp up Germany's energy goals, including the commitment that the German economy be 'almost completely decarbonised' by 2050, with renewables as the main source' [94].

Assessments of the success and importance of the *Energiewende* vary widely, with conservative commentators predictably critical of its expense and of the impact of the Feed in Tariff on economic growth and profitability. Commentators with a stronger commitment to emissions reduction targets focus more strongly on Germany's remarkable progress in expanding the share of renewable electricity from 6% in 2000 to 32% in 2016 [95].

Renewables have now overtaken coal as Germany's most important power source with solar panels and wind turbines now supplying up to half the country's electricity demand on sunny and windy days. 334,000 workers are employed in the renewable energy and energy-efficient industries [96]. Citizen support has remained strong with a 2015 survey confirming 80–90% support for the *Energiewende* [97]. One reason for this continuing support is the extensive participation by German citizens and local communities in the *Energiewende*, with almost half the investment in renewables being made by individual households, communities and cooperatives.

In 2019, Germany still faced great challenges in achieving the 2050 *Energiewende* goals. Emissions were not falling at the required rate with progress slower than had been hoped in reducing transport, industry and household energy use and emissions. The largest challenge of all remained winning broad public support for the swift and equitable phase out of brown coal mining and energy production.

The *German Coal Commission* was established in March 2018 with an ambitious eight-month deadline for recommending an agreed date for closing all coal-fired power stations. The *Coal Commission's* 28 members included two former Prime Ministers as well as senior representatives from energy utilities and business; trade unions; environmental NGOs; research institutions and coal-dependent regions. The Commission's final report, released on 26 January 2019 recommended the closure of all coal-fired power stations by 2038 with $40 billion committed to funding a just and orderly transition process.

Reactions to the *Coal Commission* recommendations were predictably mixed. German Green Party leader, Annalena Baerbock noted that 'as an industrialised country, Germany will end the use of coal. This would have been unthinkable without the pressure exerted by environmental and climate action groups as well as by the Greens. Thank you all! But we'll need more than what the compromise says if we want to meet the climate targets' [98]. Michael Vassiliadis, head of the German miners' union argued that 'we have found a compromise that cannot make us happy, but leaves us overall satisfied. We managed to shield the employees in coal power generation from social hardships from the structural change. The regions get money for structural change, to create new quality jobs. The commission laid the foundation for a new Energiewende of reason' [98].

Martin Kaiser from *Greenpeace Germany* commented that 'Germany finally has a road map for how to make the country coal-free. There will be no further coal plants. However, the report has a grave flaw: the speed is not right. The end-date of 2038 is unacceptable. The conflict about coal and the necessary speed will only end, once the exit fits the goals of the Paris Agreement' [98]. Laura Weis, from, 350.org Germany was even more critical. 'The proposal to exit coal only in 2038 is disastrous from a climate policy point of view. It's incomprehensible that the exit should only happen in 2038 after billions of euros will have gone to coal companies, industry and the mining regions. We expect that the government steps up efforts and decides an exit path that is compatible with limiting global warming to 1.5 degrees' [98].

The broader significance of the Energiewende and of *German Coal Commission* is perhaps best captured in the following quote from Johan Rockström, *Director of the Potsdam Institute for Climate Impact Research*.

> The whole world is watching how Germany, a nation based on industry and engineering, the fourth largest economy on our planet, is taking the historic decision of phasing out coal. This could cascade globally, locking in the fastest energy transition in history. This can help end the age of finger-pointing, the age of too many governments saying: why should we act, if others don't? Germany is acting, even if the commission's decision is not flawless….To avoid crucial elements of the Earth system to tip, such as the huge ice sheets, we need social tipping points. [98]

In 2019, Rockström coauthored a highly influential article on 'Social tipping dynamics for stabilizing Earth's climate by 2050' providing an evidence-based analysis of social tipping points with the strongest potential to further accelerate decarbonisation of planetary socio-economic systems. The six most promising social tipping points, informed by an extensive literature review and survey of many of the worlds' most eminent social, economic and climate scientists were identified as (i) removing fossil fuel subsidies and incentivising decentralised energy generation; (ii) accelerating the construction of carbon neutral cities; (iii) divesting from assets linked to fossil fuels; (iv) broadening public consensus about the ethical and moral implications of ongoing reliance on fossil fuels; (v) strengthening climate education and citizen engagement; and (vi) improving the transparency of information about greenhouse gas emissions trends and drivers [99].

The Great Society, the Moon Shot, the Civil Rights Movement of Our Generation: The *Sunrise Movement* and the *Green New Deal*

On Monday 13 November 2018, 200 young activists from the *Sunrise Movement* occupied the office of Nancy Pelosi, incoming speaker of the US Congress. The activists who were joined by newly elected New York Congresswomen Alexandria Ocasio-Cortez were calling for the establishment of a Congressional Select Committee to draft detailed plans for a Green New Deal, a comprehensive plan to achieve the swift and just decarbonisation of the US economy [100]. Co-founder Varshini Prakash explains the ways in which the *Sunrise Movement* builds on learning about the limitations of earlier forms of climate activism.

> We were young people from across the youth-climate sphere, people who were leading delegations of young people to the U.N. climate talks, people who had started some of the fossil-fuel divestment campaigns on college campuses in the country. All of us were feeling excited about our work and the victories we had, and at the same time feeling terrifying unease that the hurricanes were getting bigger, the fires were getting more powerful, but that our movement wasn't growing alongside that. So we launched Sunrise as an organization to move thousands of young people to participate in the political process and to push for the real solutions, at scale, that we need. [101]

The *Green New Deal* proposal championed by *Sunrise* is built on three linked goals: decarbonising all sectors of the economy at emergency speed; creating high-quality jobs through a massive expansion in the public investment; and building a just transition to support and protect the communities most affected by climate change, pollution and other environmental harm. Varshini Prakash notes that 'the *Green New Deal* is an umbrella term for a set of policies and programs that will rapidly decarbonize our economy, get all of us off of fossil fuels, and work to stop the climate crisis in the next 10 to 12 years.'

> This is a project that helps us get to 100 percent renewable energy, fast. A project to create tens of millions of jobs for working people. And we're going to fight for these jobs to be good-paying jobs, with the right to collectively bargain and include family-sustaining wages and benefits. We're pushing for the *Green New Deal* to massively invest in communities that have been on the front lines of poverty for decades and communities that historically have been left out of the political process — communities of color and low-income communities. At its core, it's a program designed to stop the climate crisis and fight poverty at the same time. [101]

Achievement of these goals will be, as Alexandria Ortasio-Cortez fully recognises, no small task. 'This is going to be the Great Society, the moon shot, the civil rights movement of our generation. That is the scale of the ambition that this movement is going to require' [102].

The strategy for implementing the *Green New Deal* included several steps. The first goal was to win support from the Democratic Caucus to establish a Congressional Select Committee empowered to draft legislation for a comprehensive plan that would fully decarbonise the economy while generating millions of jobs and reducing inequalities. One central, controversial demand was the requirement that Committee members should only include representatives who received no funding from the fossil fuel industry. By setting a reporting deadline of early 2020, the aim was for support for the draft Climate Emergency legislation to become the central, defining issue of the 2020 Presidential and Congressional elections.

While the *Green New Deal* proposal quickly attracted broad public interest and in principle support from leading Democrats, the Democratic leadership in fact approved a Climate Action Committee with guidelines far more vague than the proposed *Green New Deal Committee*. This prevarication did not come as a great surprise to *Sunrise Movement* organisers such as Evan Webber who proceeded to outline Plan B. 'If they're not going to develop the plan we will. We'll get together the scientists, the engineers, the community leaders, the mayors and city councillors, create the plan ourselves, and go out and build the public and political support to make it happen over the next two years' [103].

While the jury is still out on the long-term impact of their actions, the passion and determination of the *Sunrise Movement* and other advocates of the *Green New Deal* have already opened up space for a far broader debate about the policies needed to further accelerate climate action within and beyond the United States. And as Ocasio-Cortez argues the debate over the Select Committee is only the first skirmish in a much larger battle. 'Our ultimate end goal isn't a Select Committee. Our goal is to treat Climate Change like the serious, existential threat it is by drafting an ambitious solution on the scale necessary – aka a *Green New Deal* – to get it done' [104].

By the beginning of 2021, as the COVID-19 pandemic ripped across Europe and America and as Joe Biden emerged victorious from the US Presidential election, *Green New Deal* ideas continued to galvanise global debates about the key role which climate action and energy transition investment could play in revitalising post-pandemic employment opportunities.

I Want You to Act Is if Your House Is on Fire. Because It Is: *School Strike for Climate*

On 9 September 2018, 15-year-old Swedish school student Greta Thunberg sat down on the steps of the Swedish Parliament to begin her strike from school in protest against the lack of climate action by the Swedish Government. Thunberg's concern about climate change began when she was nine.

> The first time I heard about global warming, I thought: that can't be right, no way there is something serious enough to threaten our very existence. Because otherwise, we would not be talking about anything else. As soon as you turn on the TV, everything would be about this issue. Headlines, radio, newspapers. You would never read or hear about anything else. As if a world war was raging. But in fact, no one was talking. [105]

On 5 December 2018, Thunberg rose to address the delegates at COP 24 in Katowice Poland. Her message was short and sharp. She had come to tell them that the time for pleading for action had passed. This is now, she argued, a time of consequences, a time for responsibility, a time for action.

> We have not come here to beg the world leaders to care for our future. They have ignored us in the past and they will ignore us again. We have come here to let them know that change is coming whether they like it or not. The people will rise to the challenge. And since our leaders are behaving like children, we will have to take the responsibility they should have taken long ago. [106]

Within a few months Greta Thunberg's strike action had triggered an international *School Strike 4 Climate* movement bringing together school students in many countries demanding their governments accelerate decisive, emergency speed climate change action.

In October 2018, two 14-year-old students in Castlemaine (a small Australian town one hour drive north of Melbourne) Harriet O'Shea Carre and Milou Albrecht were inspired by Greta's example to lead a delegation of students protesting outside the office of the local member of Parliament. 'The politicians aren't listening to us when we try to ask nicely for what we want and for what we need,' said Harriet. 'So now we have to go to extreme lengths and miss out on school.' Milou added, 'Greta, one girl going on strike for a few days, or a few weeks, has made headlines. Now that someone has taken the initiative and come up with an idea that could help solve this, everyone has just jumped into it' [107].

On 27 November 2018 hundreds of students marched to the Australian Parliament in Canberra, demanding sharp acceleration in the speed of climate action. While the Conservative Australian government remained unmoved, Labor, Green and Independent Senators passed a motion expressing support for the striking students. By 29 November, the Australian *School Climate Strike* movement had spread across the country triggering school strike action and demonstrations by thousands of students in every capital city as well as many regional towns and communities. The students expressed their views in the following way.

> We are striking from school to tell our politicians to take our futures seriously and treat climate change for what it is – a crisis. Simply going to school isn't doing anything about climate change. And it doesn't seem that our politicians are doing anything, or at least not enough, about climate change either. So, as our contribution to the changes we want to see, we are striking from school. We are temporarily sacrificing our education in order to save our futures from climate wrecking projects like the Adani coal mine. [108]

In Sydney, 14-year-old Jean Hinchliffe told her fellow students the protest was just getting started. 'This is our first strike,' she said. 'Our first action. And it is just the beginning. And we'll keep doing it until something is done' [109]. Lucie Atkin-Bolton, the eleven-year-old school captain of Forest Lodge public school, said she had been let down by politicians. 'I'm the school captain at my primary school. We've been taught what it means to be a leader. You have to think about other people. When kids make a mess, adults tell us to clean it up and that's fair. But when our leaders make a mess, they're leaving it to us to clean up' [109].

The *School Climate Strike* movement continued to grow at extraordinary pace over the next 12 months with over 4 million people from 185 countries joining the Global Climate Strike on 19 September 2019 [110]. In Kiribati student protestors chanted 'we are not sinking we are fighting.' In Sydney, high school student Danielle Porepilliasana, one of over 300,000 Australian demonstrators, responded sharply to politicians telling her she should be at school. 'World leaders from everywhere are telling us that students need to be at school doing work. I'd like to see them at their parliaments doing their jobs for once' [110].

In Delhi, the demonstrators focused on the unbreathable air with 18-year-old, Rishika Singh pointing to the stark reality that 'it's the poorest who suffer the most. The rich are better off – they make the use of air conditioning and private cars for comfort.' African students targeted the offices of energy and chemical companies. German protestors blockaded the officers of the Chancellor. In the United States, thousands of Amazon, Twitter, Facebook and Google workers walked off the job to join the strike. The University of California chose this day of student action to announce it was divesting $80 billion in fossil fuel shares. And in Battery Park New York hundreds of thousands of young demonstrators roared their approval as Thunberg spoke witheringly about famous politicians seeking selfies with her. 'They tell us they really, really admire what we do yet have done nothing to address the climate crisis. We demand a safe future. Is that really too much to ask?' [111]

As news and images of these events continued to flow in I was struck like many others by the realisation that this enormous global mobilisation was unfolding just 12 months after Greta Thunberg first sat down on the steps of the Swedish Parliament. If I had told anyone in late 2018 of the speed with which the *School Climate Strike* movement would grow, I doubt if even the most optimistic climate activist would have believed me. They would probably also have been very doubtful if I had suggested that another even more radical climate action movement called *Extinction Rebellion* would have spread so rapidly around the world. Or that the term *Climate Emergency*, once seen as wildly alarmist would become so widely accepted that it would be announced as the *Oxford Dictionary's* 2019 *Word of the Year*.

These seven brief stories are just a few bright glimpses of the vast array of climate actions sweeping across the globe in the first few decades of the twenty-first century. I'm confident many readers will also have their own shortlist of campaigns and projects which most vividly illustrate the multitude of ways in which collective resistance, imagination and tenacity can tip the course of history in startlingly unexpected directions.

In my country, Australia I could, for example, point to thousands of projects created and sustained by millions of people who refuse to be blinded by denial or paralysed by despair. Climate emergency advocacy groups like the *Australian Youth Climate Coalition, Climate for Change,*

Extinction Rebellion, Climate Reality and *Farmers for Climate Action*. Policy and research engine rooms like *Beyond Zero Emissions*, the *Climate and Health Alliance, Climate Works, Solar Citizens, The Next Economy* and *The Australia Institute* demonstrating the massive potential for well-managed zero-carbon economy transitions to improve health and well-being and create millions of new jobs. *Market Forces, Environmental Justice Australia* and the *Australian Centre for Corporate Responsibility* empowering citizens and investors to hold financial institutions and businesses accountable for the global warming and environmental consequences of their investment decisions. Arts and cultural networks like *Climarte, Playback Theatre and Green Music Australia* illuminating and celebrating the beauty of the world while catalysing the transformational social and cultural changes required to overcome the threats and dangers looming into view [112].

And crucially too, in Australia, the work of Indigenous climate justice networks like the *SEED Indigenous Youth Climate Network* [113]. Recognising the many ways in which climate change risk and damage intensifies the consequences of centuries of colonial violence and respecting the depth of Indigenous learning about survival and resilience; trust and interdependence are two compelling arguments for grounding this journey of ideas in learning and wisdom from Indigenous and First Nation communities.

4

Caring for Country: Indigenous and First Nations Learning About Survival, Resilience and Resistance

Decisive, just, emergency speed actions to reduce emissions, draw down CO_2 and strengthen resilience are all essential priorities for sustaining a world in which human beings and all other species continue to thrive and flourish. At the end of the working day, in the stillness of the evening the activists and artists, researchers and writers, entrepreneurs and policy makers leading these actions will also sometimes share the fears and spectres that appear in the quiet darkness of the night. The following chapters therefore explore a range of broader conversations about insights and wisdom from diverse philosophical, theoretical and spiritual traditions which I and others who I know and respect have found helpful in sustaining meaning and purpose in the face of mounting evidence and stories and images of damaged, drowned and burning worlds.

My decision to begin this journey with a series of reflections on learning from four Indigenous and First Nations climate change writers whose work I have found particularly insightful is predicated on respectful acknowledgement of the traditional owners and elders of the country on which I have been fortunate to be born. It is also informed by increasingly sharp awareness of my responsibilities as a descendant

© The Author(s), under exclusive license to Springer Nature
Switzerland AG 2021
J. Wiseman, *Hope and Courage in the Climate Crisis*,
https://doi.org/10.1007/978-3-030-70743-9_4

of white colonial settlers born in Melbourne on the stolen and unceded lands of the Kulin Nation of south-east Australia.

Wiseman's Ferry Inn which stands on a rise overlooking the Hawkesbury River north of Sydney was built in 1821 by my convict ancestor Solomon Wiseman. My parents told me many colourful stories about Solomon's journey from convicted London timber smuggler to successful Australian landowner. However, it was only when reading *The Secret River*, Kate Grenville's novel, inspired by Wiseman's life that I began to reflect more carefully on the implications of being the privileged beneficiary of several centuries of European occupation of a country settled by Indigenous Australians for many thousands of years [114].

Some years later my aunt, Judith Wiseman gave me a copy of *Thomson Time*, a book she was proud to co-author with the eminent Australian anthropologist Donald Thompson [115]. Thomson's photos documenting the traditional way of life of the Yolngu people of Arnhem Land went on to inspire Rolph De Heer's wonderful film *Ten Canoes*. Watching *Ten Canoes* was another step for me in recognising how much I had to learn from Indigenous Australians, tempered also by awareness of the need to avoid romanticising and appropriating Indigenous knowledge.

Initial drafts of this chapter were written over the first few weeks of 2020 as bushfires engulfed vast areas of south-east Australia, threatening and destroying thousands of communities, homes and farms and killing millions of animals, birds and fish. Residents of Melbourne and Sydney rushed to buy gas masks to protect pregnant women, children and old people from the acrid smoke choking their cities. Terrifying images of towering walls of flames filled Instagram, Twitter and YouTube news feeds all around the world.

Loreena Allam, an Indigenous women holidaying at that time with her family on the south coast of New South Wales provides the following reflections on the deep wounds reopened by these events and images. 'Like you' Allam reflected, 'I've watched in anguish and horror as fire lays waste to precious Yuin land, taking everything with it – lives, homes, animals, tree – but for First Nations people it is also burning up our memories, our sacred places, all the things which make us who we are' [116].

It's a particular grief, to lose forever what connects you to a place in the landscape. Our ancestors felt it, our elders felt it, and now we are feeling it all over again as we watch how the mistreatment and neglect of our land and waters for generations and the pig-headed foolishness of coal obsessed climate change denialists turn everything and everyone to ash. Maybe this summer is the turning point, where our collective grief turns to action and we recognise the knowledge that First Nations people want to share, to make sure these horrors are never repeated. [116]

If We Fail to Care for Country, It Cannot Care for Us

Australian Indigenous author Tony Birch has written extensively and eloquently about the violence and suffering experienced by Indigenous people as a result of colonial invasion and oppression. Birch's writing draws on his rich Indigenous, Barbadian, Irish and Afghani heritage to illuminate the wisdom and tenacity enabling Indigenous, immigrant and working-class communities to survive and thrive through even the darkest, toughest times.

In recent years, Birch has also focused energy and attention on the struggle for climate justice and on creating respectful, ethical conversations between Indigenous and non-Indigenous Australians about the learning needed to rise above climate change denial and despair. 'We need,' Birch argues 'to find new ways and places to talk, to give recognition to the wealth of knowledge of climate, local ecologies and the environment more generally held and practised within particular Indigenous communities' [117].

Birch's search for ways of facilitating more productive and ethical dialogue has been strongly influenced by the work of Australian anthropologist Deborah Bird Rose on the value, risks and challenges of genuinely open and respectful cross-cultural conversations. Rose's long experience of living and working with Australian Indigenous communities led to her the view that 'to be open is to hold one's self available to others: one takes risks and becomes vulnerable. But this is also a fertile stance: one's own ground can become destabilized. In open dialogue one holds oneself available to be surprised, to be challenged and be changed' [118, p. 22].

Rose also notes the importance of recognising 'the risks at stake for Indigenous people entering into these conversations. Not only are Indigenous communities subject to knowledge appropriation, the concept of inclusion itself, however meaningful, can be debilitating and disempowering for Indigenous people' [117]. Birch builds on these observations to outline several essential preconditions for creating ethical and constructive dialogue between Indigenous and non-Indigenous cultures about justice and resilience in harsh climates and damaged ecologies.

The first, crucial step, Birch argues is full recognition of and reparation for the many ways in which the legacies of colonial invasion and dispossession continue to fuel racism, violence and inequality. Successfully addressing this challenge depends on decisive action to overcome the selective amnesia enabling many generations of settlers to avoid facing harsh realities about the sources and consequences of our continuing prosperity. Hopefully some of the insights and strategies required to overcome this wilful ignorance and blindness can also be useful in healing and curing the equally widespread disease of climate change denial.

Post-colonial farming, forestry and mining practices have all been predicated on the theft and appropriation of Indigenous land. These practices have also had a wide range of problematic consequences for Australia's soil and water resources as well as for the diversity and sustainability of many Indigenous plants and animals. Social and spiritual impacts from the physical displacement of Indigenous people are therefore often further accentuated by damage to the fragile and complex ecologies of the lands on which many Indigenous people continue to live. The destructive impacts of climate change intensified fires, floods and droughts on Indigenous lands and ecologies also further illustrate the ways in which communities which have played the smallest role in producing fossil fuel emissions are often the ones suffering the worst consequences.

Many Indigenous writers and activists remain understandably suspicious of the argument that Indigenous people should be expected to 'rise above' the ongoing violence and inequality of colonial settler legacies in order to face the 'common enemy' of climate emergency. Others, including Birch, take the view that the shared threat of catastrophic climate change, to all other species as well as to human beings, is now so acute that all potentially effective avenues for collaborative, collective action should be carefully explored. Without in any way underestimating the need to maintain a sharp focus on post-colonial violence and injustice, Birch is therefore keen to test the possibilities of more open and collaborative climate action conversations between Indigenous and non-Indigenous communities.

The principles and practices of environmental care and management employed by Australian Indigenous communities, often expressed through the idea of 'caring for country' provide one useful starting point for addressing the question identified by Birch as the greatest challenge facing non-Indigenous communities. 'How does the nation move from a state of colonial anxiety that refuses genuine recognition and engagement to a concept of locating Indigenous theories, methodologies, and methods at the centre, not the periphery of our society?' [119]

In my understanding, 'caring for country' means paying far closer attention to the complexity and fragility of the particular environments and ecologies in which we live. It means strengthening awareness that short-term choices about the resources we consume and the waste we leave behind have long-term consequences. It means giving careful consideration to the ongoing relevance of Indigenous knowledge about food and water; fire and forest management. And, as Birch notes in talking about his work in engaging young people from diverse cultures in climate change conversations, it means learning the power of climate justice story-telling, grounded in the personal experience of the places we love.

> We discussed the relationship between climate change and the havoc created by 'natural' disasters, including hurricanes, floods and ferocious bushfires such as the Black Saturday fires that devastated my home state of Victoria in 2009 and killed 173 people. I talked about country in the sense that Indigenous communities in Australia understand and experience it. The students agreed that we must listen to those who have lived

with country for thousands of years without killing it, and in order to live with a healthy planet we need to tell stories of our experience with it, and our love for it.

Finally, I asked each student a question: 'What are we seeking when we speak of climate justice?' The universal response was not restricted to justice for humans alone. My students had come to believe that if we fail to care for country, it cannot care for us. [120, p. 208]

If You Don't Move with the Land, the Land Will Move You

Tyson Yunkaporta, author of *Sand Talk: How Indigenous Thinking Can Save the World* is a member of the Apalech Clan of Queensland, Northern Australia [121]. Yunkaporta uses the term 'umpan,' derived from the word the Apalech community uses for cutting, making and writing to describe the rich mix of yarning and story-telling; sand drawing and wood carving, poetry and prose he employs to build bridges between Indigenous and non-Indigenous ways of being and understanding.

Yunkaporta notes that one of the key differences between Indigenous and non-Indigenous perspectives is their approach to context and complexity. 'Indigenous thought is highly contextualised and situated in dynamic relationships with people and landscape, considering many variables at once. Non-Indigenous thinking is good at examining things intensively in isolation' [122]. Attempting to briefly summarise the intricate and nuanced patterns, pathways and landscapes explored in *Sand Talk* sounds therefore like a rather unwise project. My more modest aim here is to bring to the table just two of the many *Sand Talk* stories which I find particularly provocative.

Anyone who has encountered the ungainly, disruptive, sometimes intimidating behaviour of Australia's largest flightless bird, will recognise Yunkaporta's account of the mess created by the mythological Emu, 'running around showing off his speed and claiming his superiority, demanding to be boss and shouting over everyone....' [121, p. 31].

Emu is a troublemaker who brings into being the most destructive idea in existence: 'I am greater than you; you are less than me....You can see the dark shape of Emu in the Milky Way. Kangaroo (his head the Southern Cross) is holding him down, Echidna is grasping him from behind, and the great Serpent is coiled around his legs. Containing the excesses of malignant narcissists is a team effort. [121, p. 30]

Malignant, narrowly short-sighted narcissism is, Yunkaporta suggests, a pervasive human failing with increasingly lethal consequences. 'All Lawbreaking comes from placing yourself above the land or above other people.' Many traditional Aboriginal customs and practices have therefore been designed to limit destructive behaviour of this kind by teaching and embedding values of respect and care for other human beings and other species as well as for the land and water on which we all depend. The problem we face now in a fragmenting, globalising world where narcissistic individualism has become the core assumption of dominant economic and political paradigms is that too often 'we have to deal with these crazy people alone, as individuals butting heads with narcissists in a lawless void....Engaging with them alone is futile — never wrestle a pig, as the old saying goes; you both end up covered in shit, and the pig likes it' [121, p. 31].

The history of Homo sapiens includes numerous examples of the speed with which social, economic and ecological crises can descend into lawless, xenophobic violence. Growing evidence of the links between climate crisis, drought and famine; the closing of borders and the building of walls suggests we would be wise to pay close attention to learning about social practices and ways of life necessary to prevent a rapid slide into barbarism. Such learning, Yunkaporta argues, might usefully be informed by values and practices which strengthen awareness of the importance of connectedness, diversity and reciprocity and by aligning our economies and ways of life more closely with the land and ecologies of the places where we live.

In the story of the Moon Sisters Yunkaporta introduces us to two sisters who long ago mistook the reflection of the moon on the ocean for a beautiful white fish. No matter how long they chased the fish or how many spears they threw, the fish continued to elude them. The shadows of the two sisters still visible on the face of the Moon are a useful illustration, Yunkaporta suggests, of the danger of seeking swift and simple conclusions about the nature of reality in a world of constantly shifting patterns and relationships.

> There can be no exchange or dialogue until the protocols of establishing relationships have taken place. Who are you? Where are you from? Where are you going? What is your true purpose here? Where does the knowledge you carry come from and who shared it with you? What are the applications and potential impacts of this knowledge on this place? What impacts has it had on other places? What other knowledge is it related to? Who are you to be saying these things. [121, p. 169]

Questions of this kind might usefully inform non-Indigenous knowledge about climate change impacts and responses. They also draw attention to the importance of broadening the ways in which we think about complex ecological relationships and risks. Increased awareness of context, patterning and complexity can strengthen support for respectful conversations drawing on diverse sources of knowledge and in building support for the difficult trade-offs required to navigate just and resilient zero-carbon transitions.

Mumma Doris, one of many Indigenous elders who Yunkaporta yarns with in *Sand Talk* speaks, for example of the principles of 'Respect, Connect, Reflect and Direct,' implemented in that order, as four key steps in building broad support for transformational cultural, social and economic change.

> The first step of Respect is aligned with values and protocols of introduction, setting rules and boundaries....The second step, Connect, is about establishing strong relationships and routines of exchange that are equal for all involved....The third step, Reflect, is about thinking as part of the group and collectively establishing a shared body of knowledge to inform what you will do.....The final step, Direct, is about acting on that shared knowledge in ways that are negotiated by all. [121, p. 275]

Mumma Doris also notes the disastrous consequences resulting from the common, ongoing tendency for non-Indigenous policy makers to intervene in Indigenous communities by working through these steps in the wrong order. Begin by telling people what to do. When this doesn't work attempt to fix the problem through consultation and recalibration. As a last resort try to repair the damage by belatedly acknowledging the importance of respectful dialogue and shared problem-solving. Unfortunately, it is also far too easy to think of many occasions on which such hierarchical, top-down decision-making processes have fatally undermined well-intentioned climate change mitigation and adaptation strategies.

Learning from other cultures, particularly those with strong oral storytelling and knowledge sharing traditions can help maintain a healthy sense of humility and perspective about the value of paying attention to diverse sources of evidence about climatic and ecological cycles, risks and tipping points. Australian Indigenous oral histories are full of stories about the disastrous consequences of communities failing to recognise and adapt swiftly enough to climatic and ecological change. 'All over Australia' Yunkaporta notes 'we have stories of past armageddons, warning against the behaviours that make these difficult to survive and offering a blueprint for transitional ways of being, so that our custodial species can continue to keep creation in motion.....' [121, pp. 78–79].

> Move with the land. Maintain diverse languages, cultures and systems that reflect the ecosystems of the shifting landscapes you inhabit over time....Learn from the coping strategies and resilience of refugees and work with them to design ways of living well in periods of displacement....Learn from other displaced communities who are struggling to survive at the margins, through dialogue and caring for each other.' [121, p. 79]

Climate Justice at Emergency Speed

Indigenous environmental activist, researcher and author Kyle Whyte is Professor of Philosophy and Community Sustainability at *Michigan State University* and an enrolled member of the *Citizen Potawatomi Nation*.

The *Potawatomi Nation* are a clan of the Algonquian speaking Anishinaabe people, driven south from Lake Michigan by French and British settlers during the eighteenth century.

On 30 August 1838 the Potawatomi community at Twin Lakes, Indiana was attacked and destroyed by soldiers under the command of General John Tipton. Tipton's soldiers, acting on instructions from the Governor burned all the village crops and dwellings and imprisoned the tribal chiefs. Over the next 60 days, the Potawatomi were force marched at gunpoint and under a blazing summer sun 1000 kilometers south west to Osawatomie, Kansas. The Potawatomi remember this march as *The Trail of Death*.

Indigenous communities all around the world, Kyle Whyte notes, share similar memories and histories of the catastrophic social, cultural and ecological consequences of colonial invasion. 'Colonialism is itself a form of anthropogenic climate change. U.S. settler colonialism….in a short period of time, inflicted displacement, drastic ecological changes, and lost or disrupted relationships with hundreds of species that indigenous peoples depended on through kinship ties for generations' [123]. Increased understanding of Indigenous resistance and resilience may therefore, Whyte argues, have significant implications for the ways in which Indigenous and non-Indigenous people endure and survive the climate emergency now unfolding around us.

For Indigenous peoples Whyte's analysis strengthens understanding of strategies required to identify and resist the ongoing risks and harms of settler colonisation. 'Settlement is deeply harmful and risk-laden for Indigenous peoples because settlers are literally seeking to erase Indigenous economies, cultures, and political organizations for the sake of establishing their own. Settler colonialism, then, is a type of injustice driven by settlers' desire, conscious and tacit, to erase Indigenous peoples and to erase or legitimate settlers' causation of such domination' [124, p. 135].

The Anishinaabe experience of sustaining resilience in times of disruptive displacement and migration is informed by the understanding that 'relationships of interdependence and systems of responsibility are not grounded on stable or static relationships with the environment' [124,

p. 129]. The key challenge therefore is to discover and implement strategies for sustaining cultural and social continuity in the face of constantly shifting, rapidly deteriorating political, cultural and ecological contexts and circumstances.

The first implication of Whyte's argument for non-Indigenous peoples is the need to fully acknowledge and address the many ways in which climate change impacts intensify ongoing legacies of colonial violence and displacement. Building on a strong foundation of climate justice and informed by careful awareness of the risks of New Age romanticism, we might also usefully pay closer attention to Potawatomi and Anishinaabe learning about surviving and thriving during periods of disruptive climatic, ecological and social change. Drawing on values strikingly similar to principles prioritised by Indigenous Australian communities, Anishinaabe stories and traditions emphasise and encourage behaviour and action based on trust and interdependence; reciprocity and redundancy.

> Trust refers to a quality of relationships among people in the community in which each party or relative, human and nonhuman, takes to heart the best interests of the other party or relative....Interdependence highlights reciprocity or mutuality between humans and the environment as a central feature of existence....Reciprocity is understood through the gift-giving and -receiving relationship in which each party has a special contribution to make....Redundancy is having multiple options for adaptation when changes occur and for being able to guarantee sufficient opportunities for education and mentorship for community members. [124, p. 132]

Whyte argues that these 'relational qualities' are a crucial basis for well-coordinated, swift and inclusive decision making and action in times of social and ecological crisis. 'Societies with high levels of trust, strong standards of consent, and genuine expectations of reciprocity will be able to work together to ensure that forest conservation or resettlement programs can be enacted quickly and justly when they are needed. In the absence of these qualities, speediness is likely possible only if consent or reciprocity are violated' [123].

Anishinaabe traditions also place strong emphasis on the importance of reciprocity with past and future generations as well as with other species. Focusing on our responsibilities to previous generations might usefully, as Potawatomi ecologist and activist Robin Wall Kimmerer also notes, encourage us to ask questions such as: 'How do we return the gifts of our ancestors? What would they expect of us? How do we become good ancestors ourselves? In considering our accountability to future generations we might therefore choose to consider the thought experiment: What do we hope that our descendants will think of us when we who are alive now have in turn become their ancestors?' [125] Heightened awareness of our reciprocal, mutually interdependent relationships with animals and insects; birds and fish; seeds and soil; rivers and oceans; ice and snow will also become increasingly crucial if human beings and other species are to continue to survive and thrive.

Realisation that the impacts and legacies of settler colonialism, neoliberal capitalism, industrialism and consumerism may have already driven us past key tipping points in relation to both CO_2 emissions and in corrosion of support for values and practices of justice, trust and reciprocity leads, Whyte notes to two equally unsettling scenarios [123].

Scenario 1: Speed without justice. Emergency speed emission reductions actions driven and controlled by the most powerful nations and corporations keep global warming below 2 degrees. The increasingly authoritarian and undemocratic governance strategies and economic policies employed to achieve these outcomes continue to entrench inequalities of wealth and power. These strategies also further intensify the dispossession, exploitation and displacement of Indigenous communities.

Scenario 2: Justice without speed. There is significant progress in rebuilding support for values and practices of trust, accountability and reciprocity. Unfortunately, the time taken to turn around deeply entrenched, deeply unjust policies means that emission reduction outcomes are too slow to prevent global warming outcomes well above 2 degrees. This leads to increasingly destructive social, economic, health and environmental outcomes for many communities including for many Indigenous peoples. The strengthening of justice-oriented principles and practices does however create a stronger base for implementing more just and democratic adaptation and resilience responses as the climate catastrophe continues to worsen.

We Are Showered Every Day with the Gifts of the Earth

Dr Robin Wall Kimmerer, author of *Braiding Sweetgrass: Indigenous wisdom, scientific knowledge and the teachings of plants*, enrolled member of the Citizen Potawatomi Nation and Distinguished Teaching Professor of Environmental Science and Forestry at the State University of New York also describes herself as a mother, plant ecologist, storyteller and gardener [125].

Like her Potawatomi colleague Kyle Whyte, Kimmerer remains deeply affected by stories and legacies from the *Trail of Death* in which her people were violently relocated from Lake Michigan to Kansas and later further south again to Oklahoma. 'I wonder' she asks 'if they looked back for a last glimpse of the lakes, glimmering like a mirage. Did they touch the trees in remembrance as they became fewer and fewer until there was only grass...' [122, p. 13].

Kimmerer begins *Braiding Sweetgrass* by noting that the Potawatomi people speak of the land not as 'a natural resource' or as private property to be owned and exploited but as a great gift: emingoyak, that which has been given to us......'we are showered every day with the gifts of the Earth, gifts we have neither earned nor paid for: air to breathe, nurturing rain, black soil, berries and honeybees, the tree that became this page....' [126]

Failure to fully recognise and be thankful for the gift of life on Earth is, Kimmerer, argues the primary source of the catastrophic risks now facing human beings and all other species. 'Though we live in a world made of gifts, we find ourselves harnessed to institutions and an economy that relentlessly asks, "What more can we take from the Earth?" This worldview of unbridled exploitation is to my mind the greatest threat to the life that surrounds us' [125].

Our first obligation as the recipient of a gift is to appropriately and graciously express our appreciation and gratitude. This is why, Kimmerer notes the children in the First Nation school close to her home in upstate New York begin each week with the following Thankfulness Address rather than a Pledge of Allegiance to the Flag. 'We are thankful to our Mother the Earth, for she gives us everything that we need for life. She supports our feet as we walk about upon her. It gives us joy that she still continues to care for us, just as she has from the beginning of time. To our Mother, we send thanksgiving, love, and respect' [125, p. 108]. How I wonder, might we develop relevant and appropriate acknowledgements of gratitude which might resonate with children, teachers and parents in other cultural and political contexts?

In most instances, we also understand that, in accepting a gift, we are entering a relationship in which we have a responsibility to give careful thought to the gifts we will choose to reciprocate the thoughtfulness and care of the gift giver. 'We are all' Kimmerer suggests 'bound by a covenant of reciprocity: plant breath for animal breath, winter and summer, predator and prey, grass and fire, night and day, living and dying' [125, p. 383]. She asks us to consider two contrasting stories about giving and taking; gratitude and greed to help us understand the consequences of failing to honour the covenant of reciprocity between human beings and the Earth.

In one story, our grandmother invites us to her house and offers us a plate of biscuits she has made. We gratefully accept her offering, taking time to appreciate the love and care she has taken in preparing her gift to us. We return in a few days with a cake we have baked for her. Our grandmother is delighted and encourages us to return again next week for her next batch. In the second story, we hear that our grandmother has a large tin of biscuits in her cupboard. We break into the house, eat all the biscuits and steal or spoil the ingredients which would enable her to make more biscuits in the future. Our distressed and angry grandmother is unlikely to invite us back any time soon.

The implications for the choices we make in responding to the climate crisis are abundantly clear. We can continue to extract and exploit the Earth's finite resources of oil and coal and gas, further intensifying the frequency of catastrophic fires, storms and floods. Or alternatively we can deepen our appreciation of the contribution which the sustainable, renewable, freely available 'gifts of wind and sun and water' can make in accelerating the transition to a just and resilient zero-carbon economy [125].

Overcoming the Windingo: Attentiveness, Gratitude, Reciprocity and Healing

The Potawatomi, like many Indigenous peoples, teach their children a number of cautionary tales about the dire consequences of narcissistic greed and of forgetting the importance of respect and gratitude for the gifts of the Earth. 'The spring dries up, the corn doesn't grow, the animals do not return, and the legions of offended plants and animals and rivers rise up against the ones who neglected gratitude' [126].

The most terrifying of these stories involve the Windingo, a mythical man-eating monster who stalks the snow bound villages and forests of the Algonquian speaking peoples of north-eastern America. Despite his insatiable appetite for human flesh the Windingo remains 'gaunt to the point of emaciation, its desiccated skin pulled tightly over its bones. With its bones pushing out against its skin, its complexion the ash-gray of death, and its eyes pushed back deep into their sockets' [127, p. 221].

Tales of the Windingo often emphasise the beast's origins as a normal human being expelled from his village as a result of his failure to control his unrelenting selfishness and violence. Wandering starving and alone in the frozen wastes of deepest winter the monster's bitterness and anger continues to grow. Here, Kimmerer suggests another useful analogy for the sources and drivers of the climate crisis. 'The consumption-driven mindset masquerades as "quality of life" but eats us from within. It is as if we've been invited to a feast, but the table is laid with food that nourishes only emptiness, the black hole of the stomach that never fills. We have unleashed a monster' [125, p. 308].

Gratitude, attentiveness, reciprocity and regeneration are, Kimmerer argues all essential guiding principles for strengthening our capacity to meet the increasingly monstrous challenges of the climatic and ecological emergencies now unfolding around us. 'Gratitude for all the Earth has given us lends us courage to turn and face the Windingo that stalks us, to refuse to participate in an economy that destroys the beloved Earth to line the pockets of the greedy, to demand an economy that is aligned with life, not stacked against it…' [125, p. 377].

Attentiveness to the diversity and beauty; vulnerability and fragility of the Earth's gifts is, in Kimmerer's view the first, foundational step in building a culture of gratitude. Deepening our awareness and understanding of the variety and complexity of our relationships with other species can reduce the risks of ecological and evolutionary hubris.

> Paying attention to other beings — recognizing their incredible gifts of photosynthesis, nitrogen fixation, migration, metamorphosis, and communication across miles — is humbling and leads inescapably to the understanding that we are surrounded by intelligences other than our own: beings who evolved here long before we did, and who have adapted innovative, remarkable ways of being that we might emulate, through intellectual biomimicry, for sustainability. [126]

Deepening awareness of the scale and speed of mass extinctions and of the devastating impacts on other species and the environment of fires, storms and floods can also create a powerful motivation for decisive and creative restorative and regenerative action as well as for grief and despair. What actions Kimmerer asks, might we take to begin to repair the damage we have done? How can we begin to reciprocate the gifts of the Earth? This is her response.

> In gratitude, in ceremony, through acts of practical reverence and land stewardship, in fierce defence of the beings and places we love, in art, in science, in song, in gardens, in children, in ballots, in stories of renewal, in creative resistance, in how we spend our money and our precious lives, by refusing to be complicit with the forces of ecological destruction. In healing. [126]

In continuing to value and learn from Indigenous and First Nations insights about gratitude and healing; resilience and resistance I am also mindful of the need to align this learning with strong, practical support for social justice and anti-racist movements such as *Black Lives Matter* and the campaign to end *Black Deaths in Custody*. I am also keenly aware that the dominant paradigms of the world in which I have been raised derive from vastly different assumptions and narratives about the transformational power of knowledge, reason and technology.

5

Cooling the Fevered City: Reason and Hubris in Greek and Enlightenment Philosophy

In January 2019, the eminent British naturalist Sir David Attenborough made the following observations in his opening address to the *Davos World Economic Forum*.

> I was born during the Holocene—the name given to the 12,000-year period of climatic stability that allowed humans to settle, farm and create civilisations. Now in the space of one human lifetime….all that has changed. The Holocene has ended. The Garden of Eden is no more.
> It is tempting and understandable to ignore the evidence and carry on as usual or to be filled with doom and gloom. But there is also a vast potential for what we might do…..As a species we are expert problem solvers. But we haven't yet applied ourselves to this problem with the focus it requires. We can create a world with clean air and water, unlimited energy, and fish stocks that will sustain us well into the future. But to do that we need a plan. [128]

In placing such strong emphasis on the human species' capabilities as 'expert problem solvers,' Attenborough foregrounds the widely held view that rigorous scientific analysis and the reasoned application of knowledge and technology provide our last best hope for building the bridges that will lead us beyond the climate chasm. Attenborough's reference to the Holocene also leads us to a series of fiercely contested debates about the definition and implications of 'the Anthropocene,' 'the period during which human activity has been the dominant force on the climate and the environment' [129].

As a white Western male born and educated in the second half of the twentieth century, I am constantly drawn to the seductive promises of scientific and technological salvation, forged in the crucible of classical Greek and Enlightenment philosophy. If only we could harness the power and creativity of critical, evidence-based enquiry and ethically informed reason with sufficient wisdom and resolve then surely we could overcome our potentially suicidal tendencies towards short-term pleasure-seeking, denial and despair. I am drawn to this flame and then retreat, remembering all too well the mounting evidence that failure to recognise the limitations of technological wizardry and heroic faith in the inevitability of progress is precisely the kind of hubris which has fuelled the delusional pursuit of infinite growth on a finite planet.

The tension between these two perspectives underpins an increasingly intense debate between two competing points of view about the best ways of facing and addressing the climate crisis. One view, most clearly articulated in the Breakthrough Institute's *Ecological Modernisation Manifesto* is resolutely optimistic in championing the potential for 'Reason, Science, Humanism and Progress' to lead humanity into the fertile possibilities of a 'good or even great Anthropocene.'

> As scholars, scientists, campaigners, and citizens, we write with the conviction that knowledge and technology, applied with wisdom, might allow for a good, or even great, Anthropocene. A good Anthropocene demands that humans use their growing social, economic, and technological powers to make life better for people, stabilize the climate, and protect the natural world. [130, p. 6]

The authors of the *Ecological Modernisation Manifesto* are unashamedly upbeat in framing the story of the last three hundred years as an overwhelmingly positive demonstration of the potential for scientific and technological ingenuity to achieve rapid and profound improvements in human wellbeing. Look, they say, at the startling advances in economic productivity, wealth and living standards resulting from revolutionary technological breakthroughs in energy; electrification and mass production; digital, information and communication technologies; artificial intelligence, robotics and biotechnology. Or look, as Bill Gates would surely agree at the huge improvements in human health, infant mortality and life expectancy achieved as a direct result of scientifically informed knowledge about sanitation, nutrition, immunisation, medical diagnosis, anaesthesia and surgery.

The *Breakthrough Institute* recipe for keeping global warming below 1.5 degrees is disarmingly simple. Decouple economic growth from GHG emissions and accelerate the switch from fossil fuels to renewables through technological innovation and an increased price on carbon. Rebuild support for nuclear power. Accelerate investment in carbon capture and storage. And if all else fails flick the switch to the CO_2 drawdown and global cooling geoengineering technofix.

Celebration of improvements in human well-being created by scientific research and technological innovation needs however to be carefully tempered with awareness of the profoundly unequal distribution of costs and benefits. To what extent, for example, would Indigenous people in Australia, the Pacific Islands or the Amazon agree that their way of life has improved in the years following European invasion and conquest? How do triumphant claims about the upward incline of reason and progress look to hundreds of thousands of Afro-American prisoners incarcerated in the jails of the United States or to the many millions of refugees looking out through the barbed wire of internment camps in Myanmar or Australia or along the US-Mexican border?

The scorecard of costs and benefits of unconstrained economic growth and technological progress becomes even more problematic when we turn our gaze to the deeply disturbing, rapidly expanding evidence of pandemic disease, biodiversity loss, species extinctions, undrinkable water, unbreathable air—and of potentially catastrophic increases in GHG emissions.

An alternative far more sceptical assessment of the ecomodernist project is eloquently expressed by Jeremy Lent, author of *The Patterning Instinct, A Cultural History of Humanity's Search for Meaning*. 'Ecomodernism…..a neoliberal, technocratic belief that a combination of market-based solutions and technological fixes will magically resolve all ecological problems….fails however to take into account the structural drivers of overshoot: a growth-based global economy reliant on ever increasing monetization of natural resources and human activity' [131].

This bleakerand more critical perspective about the risks and consequences of human hubris also informs Paul Kingsnorth's *Dark Mountain Manifesto*.

> After a quarter century of complacency, in which we were invited to believe in bubbles that would never burst, prices that would never fall, the end of history….Hubris has been introduced to Nemesis. Now a familiar human story is being played out. It is the story of an empire corroding from within. It is the story of a people who believed, for a long time, that their actions did not have consequences. It is the story of how that people will cope with the crumbling of their own myth. [132]

The stark differences in assumptions underpinning these two points of view may help explain the frequent difficulties which climate change optimists and pessimists have in understanding each other. Many well-informed scientists, activists and policy makers, fully aware of the most confronting evidence about the severity of climate risk still hold strongly to the view that the most potent sources of hope and courage in the climate crisis lie in the uniquely human attributes of reason and logic; ingenuity and inventiveness.

Others, similarly well-informed, regard the depth of this faith in the problem-solving genius of human beings as dangerously misguided and naïve. Perhaps, they argue, a little more humility about the omniscience of human reason and the omnipotence of technological solutions might be prudent given the complexity and fragility of Earth and climate system relationships, dynamics and tipping points.

My desire to better understand the foundations and implications of this highly charged and often bitter debate about the role of rational analysis and technological innovation in addressing complex threats and challenges leads me to the following questions. Which ideas from the Western canon of Greek and Enlightenment philosophy have been most influential in shaping current debates about hope and courage in the climate emergency? Which of these insights and perspectives should we carry with us and which might be we wise to leave behind?

Critical Analysis and Rigorous Debate

The lives and work of three of the most influential philosophers of ancient Greece, Socrates, Plato and Aristotle, all overlapped with and were profoundly influenced by the devastating impacts of the Peloponnesian wars and of the great plagues which swept through Athens killing more than a third of the city's population.

The historian Thucydides describes Athens at the time of these plagues as a spiritually demoralised, nihilistic society in which law and morality had disintegrated in the face of widespread despair about the likelihood of immanent death. 'As the disaster passed all bounds, men, not knowing what was to become of them, became utterly careless of everything, whether sacred or profane. So they resolved to spend quickly and enjoy themselves, regarding their lives and riches alike as things of the day' [133, p. 152]. 'Spend quickly and enjoy yourselves, regarding your lives and riches alike as things of the day' sounds like an uncomfortably familiar brief for an advertising campaign targeting modern consumers increasingly anxious and despairing about the chasms opening up before us.

It seems likely that direct experience of life in times of war and plague contributed to the determination of Socrates, Plato and Aristotle to explore and clarify the sources of wisdom and ways of living which could strengthen the capacity of human beings to sustain meaning and purpose even in the darkest of times.

Socrates begins his observations on the conditions for living a good and meaningful life with the proposition that no human being desires what is bad for them. If individuals do something that is truly bad this must either be because they are ignorant or because they are forced to act against their will [134]. The good life, the life of wisdom, the life of virtue is therefore a life devoted to the pursuit of knowledge and the clarification of misconceptions through the Socratic method: respectful and rigorous questioning and dialogue informed by logical and critical analysis of the most robust and trustworthy evidence.

Unfortunately for Socrates, many of his fellow Athenian citizens found his passion for questioning and critique so dangerous that they sentenced him to death for 'corrupting the youth of Athens.' Socrates' refusal to make the compromises required to save his life illustrates, as theologian Paul Tillich notes, important differences between two kinds of courage: courage as the bravery of the warrior in battle and the courage we demonstrate when aligning our actions with our carefully considered core beliefs, no matter what the cost [135].

Socrates' unswerving commitment to the pursuit of knowledge through critical analysis of robust evidence remains an abiding source of inspiration for climate scientists seeking to strengthen understanding of climate change causes, trends and impacts. Socrates might also recognise and sympathise with the distress experienced by so many climate scientists continually vilified and trolled for their steadfast determination to stand by their scientific findings [46].

An overly simplistic understanding of scepticism and an unreflective obsession with critique does however bring some risks. Taken to extreme lengths the Socratic method has sometimes been used to justify deliberately misleading forms of climate scepticism and climate denial. Fossil fuel corporations have, for example, learned much from the tobacco industry playbook about strategies for undermining public support for decisive climate action by careful sowing of seeds of doubt in relation to the accuracy and validity of climate science [136].

A relentless and single-minded focus on questioning every scientific finding and doubting every course of action also has the potential to create a paralysing inability to act at the very moment at which decisive action is most urgent. The German poet Bertolt Brecht may well have had concerns like this in mind in his cautionary reflections on the relation between doubt and action [137, p. 336].

Therefore, if you praise doubt
Do not praise
The doubt which is a form of despair.
What use is the ability to doubt to a man
Who can't make up his mind?
He who is content with too few reasons
May act wrongly
But he who needs too many
Remains inactive under danger

Similar reflections on the complex, problematic relationship between doubt and despair may have led Taoist and Confucian scholar Ralph Schonfield to the proposition that 'the paradigm shift induced by climate change may well be a turn away from the Socratic model of doing philosophy that focuses on questions, cultivates doubt, and pines for wisdom as something elusive. It would be a turn toward the burden of answers and the cultivation of insight' [138].

The Just Society and the Virtues of Moderation

For Socrates' student and successor Plato, the knowledge required to sustain human happiness depends on using the power of the human mind to discover and clarify the abstract forms, ideas and principles about the good and just society which can help us understand and manage the material world of formless, constantly changing chaos.

In Plato's famous allegory of the cave, he asks us to imagine human beings as prisoners chained in darkness, watching shadows playing on the wall. Prevented by our chains from turning around, we cannot see the 'real' shapes of the forms casting these shadows. Even if a few enlightened individuals escape their chains, the prisoners left inside the cave will struggle to believe or understand the descriptions brought back to them about the reality of the forms illuminated by the sunlight in the world outside.

The allegory of the cave can be interpreted in many ways. We might, for example, see this story as a metaphor for the ways in which the cynical manipulation of information and images can be used to create fake news, moral inertia and political paralysis [139]. An alternative, less democratically minded perspective might see this story as a warning about the tendency for the poorly educated, poorly informed majority to mistrust and condemn the better educated, more rationally minded minority.

Plato appears to have had this latter view in mind in proposing that a small elite of 'philosopher kings' or 'guardians' should be charged with maintaining the rule of law necessary to sustain the good and just society. He also wonders if there may be times where these guardians might need to employ 'noble lies' and 'beautiful untruths' in order to protect the rule of law [140]. The role and power of unaccountable 'elites' in ruling and guarding the good and 'sustainable' society remain problematic. To what extent, for example should we privilege scientific and technical expertise in making tough and complex choices about the deployment and governance of various kinds of geoengineering technologies?

Philosopher Mary Lane suggests that Plato's *Republic* can also be read as a prescient diagnosis of ways in which societies can become 'diseased,' locked in a 'pathological embrace in which the dominant group imposes irrational goals on the society as a whole, and in which the satisfactions sought by the ambitious and competitive were consistently unsatisfying, leading to further degeneration as their children sought satisfaction elsewhere' [139, p. 22]. To make the implications of this analysis more tangible, imagine a society in which a small minority, perhaps one percent of the population successfully captures the vast majority of the world's resources. And imagine too the moment when the children of the disillusioned and disenfranchised majority begin to rise, creating a movement which they choose to call *Climate Strike 4 Future*....

Plato's story about the *City of Pigs* explores the ways in which the journey from the 'healthy' to the 'diseased' city might occur. The story begins with the image of a harmonious, healthy city in which all citizens contribute to the common good according to their various abilities and in return receive sufficient goods and services to meet their needs. Plato then notes the way in which insatiable desire for an ever-increasing range and quality of goods and experiences—for tastier food and finer furniture; for grander music and more exciting entertainment—leads inexorably to a more acquisitive, violent 'fevered society,' requiring ever-increasing expenditure on police and soldiers, not to mention rapid acceleration in energy consumption and greenhouse gas emissions.

Plato uses this story to illustrate the virtues of moderation and the potentially destructive consequences of seeking more than our share. Many critics of the link between consumerism and exponential economic growth have drawn on this analogy in noting the historical alignment between the sudden, rapid acceleration of greenhouse gasses associated with the industrial revolution and the successful advocacy by neo-classical economists of unrestrained consumption as a virtue rather than a vice.

Plato's pupil Aristotle sharply criticises Plato's quest for universally generalisable rules and abstract definitions of the good and just society [141]. In Aristotle's view, human happiness, human flourishing and the good society are more likely to be achieved through the application of reason: using the power of the human mind to analyse, interpret and prioritise the information we receive through our senses.

The practical wisdom to choose the best, most virtuous course of action in a specific instance depends on our capacity to make well-considered judgements based on carefully reasoned analysis of relevant and trustworthy evidence. From this perspective, the key to choosing the most rational course of action often lies in finding a well-balanced middle way between icy detachment and fiery passion.

Sound judgement and practical reason are, Aristotle argues, most likely to be achieved through habits instilled by constant practice combined with respectful, well-informed dialogue [142]. Factors which strengthen the likelihood of good decision-making processes and therefore of sound, ethically informed judgement include the availability of sufficient time and trustworthy advice; the relevance and accuracy of information and evidence; the valuing of rigorous enquiry over gossip and rumour; the capacity to look at problems from other points of view; and skilful, respectful communication. These qualities are unfortunately the precise opposite of the ways in which we might describe many examples of climate change-related policy and decision making over the last twenty years.

In reflections highly pertinent to the 'fear versus hope' debate about climate change communication strategies, Aristotle notes that fear, defined as 'pain or disturbance resulting from the imagination of impending danger' often plays a vital role in encouraging the attentiveness required for thoughtful decisions and well-considered action [143]. The opposite of fear from this point of view is confidence rather than hope. While complete hopelessness does indeed lead to despair and paralysis, an excess of confidence 'instils carelessness, negligence and heedlessness, whereas fear makes people more attentive' [144, pp. 5–6].

Philosophers Michael Lamb and Melissa Lane also emphasise the ongoing relevance for current climate change debates of Aristotle's insights on the preconditions for ethically informed collective decision making. 'By situating all communication within an ethical relationship between speaker and auditor, emphasizing the agency and judgment of auditors, and highlighting ways to build trust, Aristotle offers an art of rhetoric that can help climate scientists communicate both ethically and effectively' [145, p. 1].

Consider, for example, the complex challenges facing the person responsible for directing responses to a massive, fast-moving firestorm. Reports keep coming in of lives and houses under threat; of swirling sudden wind changes; of new fires spotting far beyond the main fire front; of panic-stricken drivers trapped in smoke and flames. Every minute brings new demands for critical and urgent decisions all with complicated and potentially catastrophic consequences. What evidence and advice should you trust in deploying your limited resources? What ethical principles and criteria should you use to weigh the risks of sending your fire crews into the heart of the fire with the risks facing the communities now under imminent threat of being overwhelmed?

The quality of the judgements you make in this maelstrom of smoke and flame; panic and confusion may depend far more on the depth of your experience in making well-informed, well-advised, well-considered decisions under fierce pressure than a guidebook of rules and procedures designed for other less unpredictable, less extreme, less volatile circumstances. On the days following the inferno, as we walk silently and reflectively through blackened forests and smouldering ruins, we may be drawn to another source of wisdom drawn from classical Greek and Roman philosophy.

Making the Best of What Is in Our Power

While Stoic philosophers such as Epictetus, Seneca and Marcus Aurelius shared Aristotle's view about the power of reason they were far more sceptical about the capacity of human beings to control the material world or the actions of other human beings. We cannot, in their view find an entirely rational, entirely satisfactory solution to every individual or societal challenge. We can however learn to live a good and well-considered life, even in the most distressing circumstances, by training our minds to focus our attention and our action on those things that can indeed be changed. 'Make the best use of what is in your power and take the rest as it happens' Epictetus advised. 'Some things are up to us and some things are not up to us' [146].

Stoicism is not, as is sometimes suggested, an argument for repressing passion or avoiding action. A truly stoic approach in fact requires us to be fully aware of our emotions but not driven by them. In rising above our anxieties and fear, we can then make conscious choices to act decisively on those issues on which we have a reasonable chance of making some positive impact. Honest, courageous, stoic responses to the climate emergency might therefore include acknowledging our distress and grief without being overwhelmed by despair; giving careful consideration to the ways in which our skills and resources can most effectively be deployed; and then proceeding to act with all possible speed.

The obvious risk in the Stoic approach is the fatalistic temptation to view the key drivers of global warming and the tragic impacts of climate change on distant regions and communities as far beyond our control. The reassuring conclusion that we do however have some influence over the factors effecting the wellbeing and resilience of the lives of our immediate family and local community may then provide a convenient justification for passivity, hedonism and survivalism.

Interest in the psychological and therapeutic implications of Stoicism has grown rapidly in the last fifty years and continues to inform and underpin the most influential psycho-therapeutic paradigm of our times, Cognitive Behavioural Therapy (CBT). The founder of CBT, Albert Ellis, an enthusiastic Stoic disciple, reportedly gave his clients the following quote from Epictetus in their initial counselling sessions. 'Men are not disturbed by things but by the view which we take of them' [147]. As noted earlier, CBT-inspired cognitive reframing techniques also form the central foundation for many of the coping strategies used by psychologists to assist in overcoming climate change-related fear and despair.

Renewed interest in Stoic ideas about the interwoven, interdependent nature of the universe, as exemplified in the following quote from Marcus Aurelius also draw on one of the two main intellectual traditions linking Greek and Enlightenment ideas to current ecological perspectives and debates. 'Constantly regard the universe as one living being, having one substance and one soul; and observe how all things have reference to one perception, the perception of this one living being; and how all things act with one movement; and how all things are the cooperating causes of all things that exist; observe too the continuous spinning of the thread and the structure of the web' [148, p. 102].

Enlightenment philosophers like Baruch Spinoza would no doubt have been broadly comfortable with this view which also aligns closely with the work of writers informed by Indigenous, Taoist, ecological and feminist ideas. The alternative, still far more influential Enlightenment perspective on the separation of human beings and nature; mind and body; subject and object, remains deeply indebted to the Platonic assumption that reality is in the end created and controlled by the human mind. Debates between these two sharply contrasting ways of making sense of the world continue to have significant implications for the ways in which we understand and respond to climate emergency threats and challenges.

Have Courage to Use Your Own Reason!

The Enlightenment—the Age of Reason—refers to the dominant intellectual movements and ideas influencing European societies and politics in the period between the seventeenth and nineteenth centuries. Among the multiple, contested interpretations of the origins of the Age of Reason is the role of the Thirty Years War (1618–1648) as a turning point and catalyst for rejection of the infallibility of the church and the divinely ordained authority of the king. Fuelled by a deadly cocktail of dynastic and religious power struggles this war led to many millions of deaths from military conflict, civilian massacres, starvation and plague.

Widespread exhaustion with the recurrent cycle of religious persecution and bloodshed created fertile ground for the disruptive provocations of Galileo (Earth orbiting the sun); Francis Bacon (the scientific method); Thomas Hobbes (rational self-interest); Descartes (the separation of mind and matter); Isaac Newton (the laws of motion and gravity); and John Locke (individual human rights) to move from marginal heresy to mainstream common sense.

The shift in dominant sources of authority from the monarchy and the church to scientific and rational analysis was also accelerated by the growth of public spaces and publications in which an increasingly affluent middle class, enriched by capitalist wealth creation and colonial plunder could engage more freely in critical questioning and debate.

For the eminent eighteenth-century German philosopher Immanuel Kant, the Enlightenment was the first time in history in which human beings had found the courage to become truly 'adult' in making 'mature' choices informed by individual rational analysis rather than by obedience to monarchical and religious authority. In Kant's view, 'Sapere aude! Have courage to use your own reason!' was the appropriately triumphant motto for the Enlightenment [149].

Historian Jonathon Israel suggests that the vast sweep of Enlightenment ideas and legacies can usefully be divided into 'mainstream' and 'radical' streams [150]. In this view, the promotion of reason and science by 'mainstream' Enlightenment philosophers like Descartes, Locke and Kant was underpinned by the goal of retaining rather than undermining respect for traditional religious and political hierarchies of morality and power. Reconciliation of the mainstream Enlightenment's apparently contradictory goals of privileging human reason while maintaining faith in a divine God relied on acceptance of two kinds of dualistic beliefs: the split between the mind and body; and the split between human beings and other species.

French seventeenth-century philosopher Rene Descartes is renowned for coining the iconic Enlightenment phrase 'I think therefore I am.' Descartes also notes, in his 1640 letter to the Faculty of Theology in

Paris, that the separation of mind and body is in fact an essential precondition for providing 'a rational as well as faith-based foundation for the acceptance of God....For although it is quite enough for us faithful ones to accept by means of faith the fact that the human soul does not perish with the body, and that God exists, it certainly does not seem possible ever to persuade infidels of any religion, indeed, we may almost say, of any moral virtue, unless, to begin with, we prove these two facts by means of the natural reason' [151].

Eighteenth-century English philosopher John Locke is often referred to as the founding father of liberalism. Locke's support however for the right to free speech extended only as far as non-conformist Protestants. Enabling Catholics to have free speech might, Locke thought provide dangerous opportunities for interference by foreign powers. Allowing atheists to speak freely was even more problematic given that 'promises, covenants, and oaths, which are the bonds of human society, can have no hold upon or sanctity for an atheist' [152, pp. 212–213].

Immanuel Kant's enthusiasm for reasoned analysis was also tempered by his view that, when playing our socially allocated role as teacher, soldier or civil servant we should be careful to obey the instructions of our superiors and confine our powers of reason to the tasks required to 'do our 'duty.' It is only when we step outside the boundaries of our professional appointments that we gain the right and perhaps the responsibility to deploy our rational and critical capabilities without constraint. This might lead us to wonder what Kant would say today about the ethical constraints and responsibilities which a senior manager in a large fossil fuel company should consider in deciding when and how to 'speak truth to power' in converting personal concerns about the environmental and financial risks of fossil fuel investment into public comment and critique.

Kant also argues that the uniquely human capacity to reason, to speak and to plan necessarily separates the species of human beings from all other species and from the natural world in which we live. 'Now in this world of ours there is only one kind of being with a causality that is teleological, i.e., directed to purposes, but also so constituted that the law in terms of which these beings must determine their purpose is presented by these very beings as unconditioned and independent of conditions in nature, and yet necessary it itself. That being is man' [153, p. 323].

Many ecological and feminist writers have noted these two cornerstone principles of the mainstream Enlightenment—the split between mind and body and the split between human beings and nature—continue to have crucial implications for understanding and responding to climatic and ecological crises. Environmental philosopher Ruth Irwin summarises the challenges which climate change presents for modern philosophers in the following way.

> The separation of subjectivity from objects that began with Plato and reached its zenith with Descartes is entirely problematic…By consolidating the narratives of the empire of man over mere things, reductionist rationality removes key constraints at the dawn of commoditization and capitalism. This world is conceived as an aggregation of material objects, meaningless in themselves and only given meaning or form by their driver. [154, pp. 40 and 72]

The assumption that the wider world of animals, fish and birds; mountains, forests and oceans only has meaning in the presence of a 'driver' either secular (human beings) or divine (God) continues to provide a comforting rationale as we note the passing of one more species or the destruction of one more ecosystem. As Jeremy Lent also notes, Enlightenment assumptions about the separation of human beings and nature continue to provide convenient justifications for a small minority of wealthy and powerful Europeans to dominate and exploit other human beings as well as other species.

> Seeing themselves as separate from nature, [Enlightenment] philosophers such as Francis Bacon led the clarion call for humankind to 'conquer nature,' while Descartes and Hobbes introduced the view of 'nature as a machine' that has dominated Western thought ever since. Europeans, driven by the credo that 'knowledge is power,' applied their newfound power to conquering, not just nature, but the inhabitants of much of the rest of the world. [155]

Ethical Action in an Interconnected Universe

The more 'radical' stream of Enlightenment thought, inspired in particular by the work of seventeenth-century Dutch philosopher Baruch Spinoza, starts from the very different premise that the universe is comprised not of separate realms of mind and body; humanity and nature but of one single substance. The language which Spinoza uses to describe this substance, 'Deus sive Natura' (God or Nature) opens the door to both pantheist and atheist interpretations as well as to radically democratic and egalitarian political ideas.

Spinoza's view of the universe as a single interconnected entity established an important foundation for philosophical, literary and aesthetic perspectives tempering faith in the omniscience and omnipotence of human reason with greater caution about the risks and consequences of anthropocentric and technocratic hubris.

Eighteenth-century German philosopher George Hegel was strongly influenced by Spinoza in warning that an unreflective obsession with instrumental rationality and of logic divorced from ethics led inexorably to the terrible technical efficiency of the guillotine: 'the coldest and meanest of all deaths with no more significance than cutting off a head of cabbage' [156, p. 590]. Some followers of Hegel might wonder whether the growing enthusiasm in some circles for geoengineering technologies as a solution to the climate crisis share a similar underlying logic.

Hegel's German compatriot Frederick Nietszche also acknowledges Spinoza's influence in alerting us to the contingent and contested nature of 'truth' and of the ways in which arguments about objective 'facts' are often used to veil and obscure individual interests and power relations. While Donald Trump seems unlikely to have studied nihilistic ideas in any depth it is interesting to reflect on the comfort he might draw from Nietszche's observation that 'truth is merely a series of metaphors or imprecise rhetorical approximations, mobilized to achieve a certain effect or a set of ends' [157, p. 171].

As Jonathon Israel also notes, the absence of a separate divine source of authority means that respectful dialogue with other human beings becomes an essential basis for constructing and governing an ethically informed society. 'If you are going to construct a moral order in the modern world what other basis do you have? If it is not the voluntaristic preferences of some divinity to be interpreted for us, then the only way we are going to come to an agreement is if we agree to consider our interests as equal' [158].

For Romantic poets such as William Blake, John Keats and William Wordsworth, the most tragic consequences of an overly narrow focus on logic and rationality lie in losing sight of other more aesthetic, less intellectualised ways of experiencing the sublime and fragile beauty of the natural world. Wordsworth's meditations on these other ways of experiencing the 'sense sublime' are famously articulated in his reflections on the ruins of Tintern Abbey.

> I have felt a presence that disturbs me with the joy
> Of elevated thoughts; a sense sublime
> Of something far more deeply interfused,
> Whose dwelling is the light of setting suns,
> And the round ocean, and the living air,
> And the blue sky, and in the mind of man:
> A motion and a spirit, that impels
> All thinking things, all objects of all thought,
> And rolls through all things. [159]

John Keats refers to key attributes of this alternative way of seeing and being in the world as 'negative capability….when man is capable of being in uncertainties, mysteries, doubts, without any irritable reaching after fact and reason' [160, p. 60]. This observation of Keats raises several questions with ongoing relevance to our current circumstances. How do we learn to live more comfortably with greater levels of uncertainty, doubt and ambiguity? What role can poetry, art and music play in helping us develop these skills?

Tensions and differences between Classical Greek, Enlightenment and Romantic views on reason and passion; truth and beauty; certainty and doubt are also reflected in the work of Mary Wollstencraft and her daughter Mary Wollstencraft Shelley. Mary Wollstencraft's *A Vindication of the Rights of Women* published in 1792 played a key role in extending Enlightenment principles of universal human rights to women. Wollstencraft' s premise, 'if the abstract rights of man will bear discussion and explanation, those of women, by a parity of reasoning, will not shrink from the same test' led logically, she argued, to the emancipatory observation that 'the divine right of husbands, like the divine right of kings, may, it is hoped, in this enlightened age, be contested without danger' [161, pp. viii and 83].

Many commentators have noted the continuing relevance of Mary Wollstencraft Shelley's novel *Frankenstein or the Modern Prometheus* for current debates about the causes and consequences of anthropogenic climate change. The images of Frankenstein which come most commonly to mind for modern readers are from the film version of actor Boris Karloff with bolts appearing from both sides of his neck. The text of the original novel tells a far more complex nuanced of the tragic consequences arising from the remarkable success of the brilliant young scientist, Dr Frankenstein in creating a living creature from inert matter.

Climate scientist Michael Wysession draws several analogies from Shelley's tragic story of the destruction unleashed by Dr Frankenstein's monster. The most obvious lesson, Wysession argues, is that 'climate change is the monster we made. We are Victor Frankenstein.' But perhaps he suggests, there is another equally relevant, equally plausible, even more disturbing interpretation.

> Victor Frankenstein makes a creature and then abandons it. In response, the creature exacts revenge on him and his friends and family. The monster gives Victor multiple opportunities to acknowledge him, but still Victor flees and denies him. By the end of the novel, there is some doubt as to who the greater 'monster' is, Victor or his creation. By analogy, humans may be the monsters for the inhumane ways in which we treat our planet. [162]

The fact that Shelley wrote her novel during her travels in Switzerland in 1816, 'the Year without Summer' provides another way of thinking about the ongoing relevance of Frankenstein. 1816 was the first of several years in which the atmospheric dust clouds created by the explosion of the Mount Tambora volcano far off in the Pacific caused bitterly cold and bleak weather all across Europe. It is possible, historian Gillen D'Arcy Wood notes, that Mary Shelley's imagining of Frankenstein's monster may therefore also have been inspired by her encounters with the many thousands of starving refugees seeking food and shelter in famine ravaged Switzerland, near to the Castle Frankenstein where she was staying [163]. D'Arcy Wood also wonders if the suffering and dislocation experienced by this earlier generation of climate refugees might bring to mind more recent experience in New Orleans and Syria; the Philippines and Central America of refugees fleeing from the bitter consequences of extreme weather events and climatic change.

Flesh, Blood and Brain, We Belong to Nature

Many ecological activists continue to see the work of Karl Marx (1818–1883) and Frederick Engels (1820–1895) as being broadly consistent with mainstream Enlightenment views about the inevitability and desirability of human mastery over nature. While support for this view has been reinforced by the environmentally destructive economic policies of nominally Communist states such as the Soviet Union and the People's Republic of China, careful reading of Marx and Engels (particularly their later work) suggests a very different interpretation. Here, for example, are Engels' reflections on the relationship between human beings and nature.

> At every step we are reminded that we by no means rule over nature like a conqueror over a foreign people, like someone standing outside nature – but that we, with flesh, blood, and brain, belong to nature, and exist in its midst, and that all our mastery of it consists in the fact that we have the advantage over all other creatures of being able to learn its laws and apply them correctly. [164]

Marxist ecological theorist, John Bellamy Foster builds on Engel's observation that 'we, with flesh, blood and brain belong to nature' in the following way [165]. If the human species does indeed 'exist in the midst of nature' then human well-being depends crucially on maintaining healthy and sustainable metabolic relationships between human beings and the natural world. The escalating speed at which climatic and ecological boundaries are being crossed provides stark evidence that a wide range of ecological relationships are now profoundly threatened. The failure of human beings to maintain sustainable ecological relationships is not just a result of poorly informed decision making, short-sightedness and greed. The 'metabolic rift' between human beings and nature is rather a direct and inevitable result of capitalist property relations.

The capitalist dynamic of constantly accelerating capital accumulation and profit maximisation leads, Bellamy Foster argues to an inexorable, endless expansion in the exploitation of natural and human resources. Infinitely expanding the exploitation of finite planetary resources drives in turn the inevitable destruction of the metabolic relationships between human beings and nature. Ashley Dawson, author of *Extinction: A Radical History*, notes that these consequences extend beyond global warming to encompass the full array of ecological risks. 'Capital must expand at an ever-increasing rate or go into crisis.' In doing so 'it commodifies more and more of the planet, stripping the world of its diversity and fecundity' [166, p. 13].

The initial work of Marx and Engels on the causes and consequences of the potentially catastrophic rift between human beings and nature was triggered and informed by evidence of declining soil fertility in England and other European countries. 'Nineteenth century agricultural scientists suggested that this trend was partially caused by a rapid shift in population from rural areas to cities, leading to reduced capacity to replenish soil nutrients from human and animal waste' [165].

Marx and Engels went on to note the broader implications of decisions by landowners and investors to maximise agricultural productivity by replacing traditional farming practices of crop rotation with a dramatic expansion in the use of organic and synthetic fertilisers. Increased demand for organic fertilisers led to fierce economic and military competition for access to guano (bird dropping) deposits in Chile and Peru. The invention of synthetic nitrogen fertiliser in the early twentieth century has in turn been a major driver of both significant improvements in agricultural productivity and in accelerating global warming. More recent examples of the potentially catastrophic consequences of 'metabolic rift' include the collapse of global fish stocks; the extinction of many thousands of species and the destruction of old growth forests [167]. Rapid expansion in CO_2 production from accelerated economic growth has also of course been the key driver of global warming.

The view that capitalist property relations and profit maximisation have been the strongest force driving ecological destruction and climate change leads many Marxist theorists to argue that 'the Anthropocene' should more accurately and appropriately be described as the 'Capitolocene.' The era we are entering is, they argue one in which the owners of capital—not all human beings—have become the dominant force on the climate and the environment. Analysis of this kind leads theorists and activists like Bellamy Foster to argue that any genuine and lasting solution to climate change requires the comprehensive dismantling of capitalism power and property relations.

> A long and continuing ecological revolution is needed—one that will necessarily occur in stages, over decades and centuries. But given the threat to the Earth as a place of human habitation—marked by climate change, ocean acidification, species extinction, loss of freshwater, deforestation, toxic pollution, and more—this transformation requires immediate reversals in the regime of accumulation.
>
> This means opposing the logic of capital, whenever and wherever it seeks to promote the 'creative destruction' of the planet. Such a reconstitution of society at large cannot be merely technological, but must transform the human metabolic relation with nature through production, and hence the whole realm of social metabolic reproduction. [168]

A 'long and continuing ecological revolution'—'system change not climate change' will in this view be an essential precondition for the radical shift in economic, political and ecological values and relationships required to set the human project on a more sustainable path. The awkward and confronting reality is however that the speed with which the climate crisis is unfolding will probably require a timetable for emissions reduction action far shorter than that required to radically transform capitalist economic, social and political relationships.

The rapidly closing window of opportunity for avoiding the most catastrophic climate tipping points raises many tough questions about the choices required to further accelerate the speed of emission reductions while also pursuing transformational changes in deeply embedded cultural values, economic relationships and political institutions. To what extent, for example should climate justice activists be prepared to forge temporary alliances with corporate and military institutions with the financial and technological resources capable of driving an emergency speed shift to a zero-carbon economy? How do we maximise the potential for human ingenuity and creativity to address wicked interlocking climatic and ecological crises while maintaining an appropriately critical perspective on the limit and dangers of technocratic silver bullets? Is it possible to construct inclusive and deliberative democratic processes consistent with the agile and decisive decision making required to meet the challenges of increasingly severe, fast moving wildfires, floods and storms? And how do we design and implement economic decarbonisation, adaptation and resilience strategies in ways which reduce rather than accentuate inequalities of income, wealth and power?

6

Illuminating the Patterns of Domination: Critical Theory, Modernity and Power

From the trenches of World War I to the breadlines of the Great Depression and from the gas chambers of Auschwitz to the firestorms of Dresden and Hiroshima the first half of the twentieth century was a tough time for anyone assuming that human progress would be an ever-ascending staircase built on reason, science and technology.

For World War I poet Wilfred Owen [169], the slaughter and chaos of the Western Front were an ominous demonstration of the monstrous capabilities of technologically enhanced weaponry. His experience of the lethal impact of long-range artillery, machine guns and mustard gas also accelerated the end of his belief in God as well as in 'the old lie: how sweet and good it is die for your country.' Deeply schooled in the pastoral, romantic poetry of Keats and Shelley, Owen had once hoped that the beauty of the natural world might replace his loss of faith in his country or in God. But that comfort too was stripped away as he watched his friends and comrades cut down on a warm Spring day in the fields of northern France [169].

> Marvelling they stood, and watched the long grass swirled
> By the May breeze, murmurous with wasp and midge....
> Hour after hour they ponder the warm field—
> And the far valley behind, where the buttercups
> Had blessed with gold their slow boots coming up....
>
> .....So, soon they topped the hill, and raced together
> Over an open stretch of herb and heather
> Exposed. And instantly the whole sky burned
> With fury against them; and soft sudden cups
> Opened in thousands for their blood; and the green slopes
> Chasmed and steepened sheer to infinite space.

Virginia Woolf, in her 1927 novel *To the Lighthouse* raises similar doubts and questions about the possibility of redemption through our experience of nature and of 'finding in solitude on the beach an answer' in a time of broken mirrors.

> That dream, of sharing, completing, of finding in solitude on the beach an answer, was then but a reflection in a mirror, and the mirror itself was but the surface glassiness which forms in quiescence when the nobler powers sleep beneath? Impatient, despairing yet loth to go (for beauty offers her lures, has her consolations), to pace the beach was impossible; contemplation was unendurable; the mirror was broken. [170, pp. 133–134]

As the consequences of the Great Depression, European colonialism and toxic nationalism continued to unfold, the skies continued to darken. Picasso's famous painting Guernica commemorates the deaths, in 1937 of 1600 civilians bombed and strafed by Hitler's air force as bombing practice during the Spanish Civil War. And Guernica, as Viktor Frankl notes was only an initial warning siren for the dreadful violence yet to come. 'Since Auschwitz we know what man is capable of. And since Hiroshima we know what is at stake' [171, p. 154]. How, in times of such negation and despair W. H. Auden asks in, *September 1 1939*, can we find sources of meaning and flashes of light strong enough to sustain us?

> Defenceless under the night
> Our world in stupor lies;
> Yet, dotted everywhere,
> Ironic points of light
> Flash out wherever the Just
> Exchange their messages:
>
> May I, composed like them
> Of Eros and of dust,
> Beleagured by the same
> Negation and despair
> Show an affirming flame [172]

The diverse responses to Auden's provocative question point us to the wide variety of maps and signposts which twentieth-century philosophers and activists; poets and artists have drawn on in seeking meaning and courage in earlier times when the path to the future has seemed at least as threatening as it does now. The following discussion focuses particularly on contributions from critical social theorists and existential philosophers.

Illuminating the Patterns of Domination

For German Frankfurt School social theorists Theodor Adorno and Max Horkheimer, writing shortly after the end of World War II, the highest priority in such bleak times is brutally honest and unflinching critique of the causes and consequences of brutal violence and oppression. 'After the catastrophes that have happened, and in view of the catastrophes to come, it would be cynical to say that a plan for a better world is manifested in history and unites it. No universal history leads from savagery to humanitarianism, but there is one leading from the slingshot to the megaton bomb' [173, pp. 319–320].

Adorno and Horkheimer identify the source of humanity's self-destructive tendencies in 'a pattern of blind domination, domination in a triple sense: the domination of nature by human beings, the domination of nature within human beings, and, in both of these forms of domination, the domination of some human beings by others' [174]. This desire for domination—of other human beings and of nature—is, they argue driven by the illusory and pathological myth that the abiding human terror of the unknown can be successfully overcome through the pursuit, at any cost, of scientific knowledge and technocratic progress.

Adorno and Horkheimer argue that full understanding of the atrocities of Fascism and the Holocaust requires us all to think and act in accordance with a 'new categorical imperative' focused on preventing any repetition of Auschwitz. They also caution sternly against naïve utopianism. Such fantasies are, in their view unlikely ever to be fulfilled given the pervasive reach and influence of capitalist economic, social and cultural relationships, including, in particular the media. Naïve utopianism also creates the risk of reopening authoritarian discourses and tendencies (with the rise of Trumpism a cautionary warning to anyone who thinks these risks every completely disappear). All that can safely and usefully be done is to continue to shine the brightest possible light on the underlying sources of power and domination, with glimmers of hope that this may provide some limited protection from the reoccurrence of barbaric extremism.

This argument has a number of significant implications for current ecological and climate challenges. Our central task now, according to this view, is to focus searing critical analysis on the power relations and vested interests driving and benefitting from the exploitation of fossil fuels and other ecological resources. It will also be important to illuminate and, where possible limit the suffering and damage caused by climate change and biodiversity loss with a particular focus on the most vulnerable communities and ecologies. We should however remain extremely wary of technocratic climate change solutions (such as geoengineering) or of radically utopian proposals for emergency speed decarbonisation which may have the (intended or unintended) consequences of accelerating the concentration of authoritarian political, corporate and military power.

The criticisms which can, in turn, be levelled at Adorno and Horkheimer's approach are similar to those made of the work of some of their other critical theory colleagues. Abstract intellectual conversations conducted in the comfortable and privileged surroundings of the university lecture hall and dining room may produce many valuable insights about the ways in which inequalities of power and wealth are created and entrenched. While these insights provide useful antidotes to naïve responses to greenwashing and wishful thinking they can also reinforce and intensify a culture of paralysis and despair [175].

Reconnecting World and Earth

While extracting clear narratives from the infamously dense prose of Adorno's Frankfurt School colleague, Martin Heidegger is no easy task (and as always it is important to begin by noting his unapologetic sympathy for Fascism), there are several threads in his work which continue to provide influential and useful contributions to discussions about meaning and action in times of climatic and ecological risk [176].

Heidegger's premise is similar to Adorno's foundational proposition that modern capitalist societies, economies and cultures are driven and framed by the addictive, illusory goal of mastering nature through technological, logical and analytic ways of knowing and acting. For Heidegger, the consequences of this technological framing are profoundly nihilistic in the ways they separate us from nature and destroy the last vestiges of our connection with older, more mystical, aesthetic and symbolic ways of being and knowing. The rediscovery of human meaning and purpose therefore relies on the rediscovery of the deep interconnections between human beings and nature; the human world and the natural Earth. The rediscovery of our connection with the Earth, Heidegger argues depends in turn on remembering that 'the human being is not the lord of beings but the shepherd of Being' and in relearning the importance of being fully grounded and 'at home' in a particular geographical location [177, p. 245].

Social theorist Gerry Gill notes that Heidegger's understanding of the split between human beings and nature in fact evolved over time from his initial formulation that 'the world and the Earth are essentially different to one another and yet never separated' to one in which the world and the Earth merge into the 'simple oneness of things' [178, p. 184].

> Heidegger articulates our desire to reject the modern scientific-technological attitude which sees the Earth as an object to be subdued, and he generates a symbolic language which evokes the possibility of a way of being in the world in which we could experience an at oneness and a caring relationship with the Earth. His conception of human history as one of sudden epochal transformations is seductive to those longing for a leap into a new age. [178, p. 181]

The seductive appeal of Heidegger's emphasis on repairing the rupture between human beings and 'Mother Earth' helps explain the ongoing influence which his ideas have had for Deep Ecology and New Age philosophers and activists. On the other hand (and noting Heidegger's own collaboration with Fascism), we also need to bear in mind the speed with which overly simplistic arguments for linking identity with place and blood and soil can open the door to populist nationalism and racist violence.

Heidegger's reflections on the complex and contested concept of 'care' provide further useful pointers to ethical and restorative ways of life in threatening and precarious times. Heidegger's view is that the action of caring and the quality of carefulness are at the heart of being alive with authenticity and meaning. As psychotherapist Rollo May suggests the tempting dream of 'carelessness' can rapidly become a nightmare. 'When we do not care, we lose our being; and care is the way back to being' [179].

Heidegger notes that we continue to use the term 'care' in a variety of ways. On some occasions, we understand 'care' as anxiety, in the sense that we are overwhelmed by our cares and concerns about ourselves, our family and friends and the wider world. At other times we use 'care' in the more positive sense of attentiveness, mindfulness and the desire to create some change for good. It would seem that many of us have both these meanings in mind when, in talking about how deeply we 'care' about climate change, we struggle to maintain a constructive balance between our determination to keep paying close attention to evidence of risk and the paralysing consequences of rising anxiety and panic.

Heidegger also draws our attention to the distinction between caring strategies driven by a desire to 'jump in' and take control over the lives of others as opposed to more genuinely respectful approaches which focus on recognising strengths and listening closely to the concerns and desires of the people we are seeking to assist. This distinction between respectful and disrespectful care, of working with rather than for others may be useful to keep in mind as we consider strategies for working to address climate change impacts on the most vulnerable and powerless individuals and communities.

The Incalculable Grace of Love

Reflecting on the chaotic uncertainty of post-war Europe and rising fear about the prospect of a world sliding back into totalitarian barbarity, Heidegger's former student, Hannah Arendt described the characteristics of the world she observed in 1950 in uncomfortably familiar terms. 'Never has our future been more unpredictable, never have we depended so much on political forces that cannot be trusted to follow the rules of common sense and self interest — forces that look like sheer insanity' [180, p. vii].

In his 2017 article *Reading Arendt in the time of climate change*, Wen Stevenson makes two valuable observations about the contemporary relevance of Hannah Arendt's reflections on the causes and consequences of totalitarianism [181]. First, and without in anyway underestimating the horror and brutality of the Holocaust, Stevenson points to the ways in which Arendt's work helps us understand the recurring tendency for the pathway to the abyss to be opened up as much by the failure of decent human beings to act as by the deliberate malevolence of totalitarian psychopaths. 'The sad truth of the matter' Arendt argues 'is that most evil is done by people who never made up their mind to be either bad or good' [182, p. 438].

The actions of corporate executives in authorising the expenditure of billions of dollars on the escalating expansion of fossil fuel extraction in full knowledge of the ecologically disastrous consequences of their decision are clearly problematic. It remains harder however, and perhaps even more urgent, to recognise and call out the preparedness of so many of us to find excuses not to act in the face of overwhelming evidence of escalating risk. How is it that I still find myself concocting superficially plausible rationalisations for avoiding looking too closely at the consequences of the flights I take and the food I eat? 'What comprehensible motive,' Stevenson asks 'could there be for poisoning the well from which one's own children must drink, much less the atmosphere itself? What kind of mindset makes one's own children and grandchildren, and everyone else's, indeed all future generations, superfluous?' [181]

There are, as noted earlier a number of distressingly familiar motives and rationalisations for tolerating actions likely to cause great harm to current and future generations. The desire to concentrate wealth and power, often conveniently justified by the dehumanisation of other human beings and the devaluing of the lives of other species. The seductive appeal of simplistic solutions to complex challenges opening the door to fanaticism and fundamentalism. The allure of short-term pleasures over long-term risks. The temptation to avoid responsibility for making hard choices. And the paralysing toxin of despair.

The urgent task of overcoming paralysing despair leads to the second provocative question addressed in Stevenson's article. How can Arendt's reflections assist us to meet the challenge she places before us of 'bearing consciously the burden that our century has placed on us — neither denying its existence nor submitting meekly to its weight?' [180, p. viii] How can we sustain the courage to keep turning up on the climate action frontline as the firestorm continues to expand? How do we continue to pay attention; to keep our eyes open; and to refuse to participate in actions and ways of life which require us to betray our core beliefs?

Arendt identifies one powerful source of sustenance and resilience in collective action; in the strength we gain from being and working together with like-minded individuals in the public realm. This observation would clearly resonate strongly with the experience of many current-day climate activists. The greatest of all comforts Arendt suggests however lies in 'the unpredictable hazards of friendship and sympathy.....the great and incalculable grace of love' [180, p. 320].

History, Arendt acknowledges 'knows many periods of dark times in which the public realm has been obscured and the world become so dubious that people have ceased to ask any more of politics than that it show due consideration for their vital interests and personal liberty'

[183, p. 11]. The future however is never entirely written. New life, new conversations, new collaborative projects create the possibility of new beginnings and new stories.

> Even in the darkest of times we have the right to expect some illumination, and that such illumination might well come less from theories and concepts than from the uncertain, flickering, and often weak light that some men and women, in their lives and their works, will kindle under almost all circumstances and shed over the time span that was given to them. [183, p. ix]

Arendt drew great comfort from her belief that the birth of every new child creates the potential for each new generation to start again; to rekindle the fire; to renew our conversations and redirect the focus of our actions. Writing in the shadow of the Cold War and the threat of nuclear annihilation, Arendt was acutely aware that there could be no guarantees about the long-term future of the human species. Her distress and concern may well have been deeper still if she could have foreseen the conversations emerging among young climate activists in the early decades of the twenty-first century about the morality of giving birth and raising children given the hostile environment which we are likely to be passing on to them [184]. I'm sure Arendt would fully understand the motivations of young people choosing not to have children at this time. She would also I imagine respect and admire the determination of many other young climate activists in refusing to let the climate crisis overwhelm their desire to bring new life into the world.

Respectful and Well-Informed Deliberative Decision Making

In 2016, the Parliament of Ireland established a Citizen's Assembly to advise on national priorities in relation to complex and divisive issues such as abortion, aged care and climate change. The Assembly was made up of 99 randomly selected citizens, broadly representative of the age, gender, social class and regional spread of all sections of Irish society. In

2018, the Assembly's deliberations played a significant role in enabling the citizens of a predominately Catholic society to conduct a respectful and constructive public debate and referendum on abortion law reform [185].

In April 2018, following a extensive discussion and deliberation chaired by a former Chief Justice of the Supreme Court, the Irish Citizen Assembly tabled its recommendations on the topic 'How the State can make Ireland a leader in tackling climate change.' Bearing in mind the contentious nature of the topic and the fact that the Assembly included a broadly cross section of Irish social and political perspectives it is impressive to note the extent of shared commitment reflected in the Assembly's key recommendations. These recommendations included 100% support for a strong leadership role for government in driving climate mitigation; 100% support for greater community ownership of renewable energy and 80% support for higher taxes to fund climate action policies [186].

As Sadhbh O'Neill, adviser to the Irish Parliamentary committee on climate change noted the work of the Assembly demonstrated the value of creating respectful, reflective dialogue in the context of a public sphere increasingly corrupted and gamed by wealthy and powerful vested interests. 'If you structure the debate around information, discussion, questions and answers, and allow citizens to really thrash things out with expert advice–in a manner that is intelligible to them – very often people will shift their positions' [186].

Positive outcomes from Citizen Assembly and Citizen Juries in many jurisdictions have encouraged climate action movements, policy makers and governments to place increasing emphasis on deliberative democracy initiatives of this kind. The 2018 *Extinction Rebellion Declaration* includes, for example, the following three core demands.

1. Government must tell the truth by declaring a climate and ecological emergency working with other institutions to communicate the urgency for change.
2. Government must act now to halt biodiversity loss and reduce greenhouse gas emissions to net zero by 2025.
3. Government must create, and be led by the decisions of, a citizen's assembly on climate and ecological justice.

The theoretical and conceptual frameworks informing these deliberative democracy practices and strategies draw heavily on the work of German social theorist Jurgen Habermas. Habermas, who succeeded Max Horkheimer as Professor of Philosophy at Frankfurt University in 1964 continues to make a major contribution to the following question central to the focus of this book. How can we find a solid, shared foundation for climate action ethical principles and priorities in ways which reach beyond the risks and limitations of faith-based revelation, subjective introspection, instrumental reason and moral relativism?

Truth and knowledge revealed to us as a result of religious faith or introspective reflection both rest, in the end on the assumption that judgements about the nature of truth and the desirability of an action can be made without reference to consequences for any other individual. The brutal history of religious and political persecution provides multiple illustrations of the dangerous slippery slope often resulting from this assumption. Instrumental reason, the use of reason as an instrument or tool for achieving social and political goals regardless of wider implications has too often been used to justify violent oppression of other individuals and communities. Moral relativism, the idea that all ethical positions are equally valid, leaves us wandering alone in a wild and arid ethical desert.

For Habermas, the most promising alternative approach to the achievement of shared ethical principles and priorities builds on the uniquely human capacity to communicate through language [187]. Habermas coins the terms 'communicative action' and 'ideal speech situations' to describe decision-making processes in which all individuals involved are able to engage in fully informed, fully respectful, conversations about the most desirable course of action. This leads him to the view that 'the only regulations and ways of acting that can claim legitimacy are those to which all who are possibly affected could assent as participants in rational discourses' [188, p. 458].

Protecting the rights of the individual to freedom of speech and freedom of association is a necessary but not sufficient condition for the effective establishment of such Ideal Speech conversations. The capacity to engage in well-informed and respectful deliberation also depends on the creation of a public sphere which enables and encourages broad and inclusive public participation in political decision making. 'Citizens behave as a public body when they confer in an unrestricted fashion – that is, with the guarantee of freedom of assembly and association and the freedom to express and publish their opinions–about matters of general interest' [188, p. 354].

Noting that the effective exercise of civil and political rights depends first of all on citizens being able to meet their basic material needs, Habermas also emphasises the importance of establishing and defending a range of 'social and ecological rights' [189]. Access to adequate food and shelter, clean water, breathable air and a livable climate are all clearly essential preconditions for creating the context within which human beings can undertake meaningful and respectful conversations on matters of shared concern.

Principles derived from Habermas's theory of communicative action have, as noted above, become an influential source of practical guidance for the democratisation of environmental and climate change decision making [189]. These deliberative democracy strategies typically begin by identifying and bringing together a comprehensive, randomly selected gathering of individuals likely to be affected by a particular decision. Once agreement has been reached on protocols for conducting the conversation and reaching decisions, the next step is to ensure all participants have access to relevant and trustworthy sources of information, evidence and analysis. This then provides the basis for a carefully facilitated process of dialogue and reflection with the aim of arriving at recommendations supported either by majority vote or consensus.

Unsurprisingly, the effectiveness of decision-making processes informed by communicative action principles has generally been greatest in addressing local and relatively small-scale environmental issues and conflicts. The task of achieving well-informed and respectful dialogue becomes considerably harder as the scale and range of issues and actors involved become larger and more complex. Habermas is also well aware of the increasing capacity of large financial and media corporations to control and manipulate the flow of information [189]. Mitigation of these risks generally depends on strong regulatory and legislative action to maintain public institutions and communication technologies which enable rather than undermine respectful and reflective public discourse.

The fact that communicative action explicitly celebrates and privileges the human capability of speech has also led some critics to raise concerns about the limitations of this approach in protecting other species and the natural world. While it is possible that a well-informed process of collective deliberation might lead to recognition of the mutual dependency of human beings and other species there is no guarantee that this will be the case.

The current course of action being pursued by Homo sapiens in fact appears to be based on the alternative premise of stripping the planet of resources in order to maximise short-term gains for the wealthiest humans while leaving future generations to deal with the consequences of species extinctions and biodiversity loss. These concerns have therefore led some ecological and political theorists to explore a range of strategies for including the 'voices' of non-human beings in deliberative democracy processes [190, p. 46].

Creativity and Beauty, Love and Friendship

One of the key questions arising from the work of writers with direct experience of Fascism and the Holocaust—and indeed of famine, plague and war and more generally—is why some individuals respond to appalling conditions with courage and others with cowardice; some with kindness and some with cruelty. This is the central question which Austrian psychoanalyst and Holocaust survivor Viktor Frankl explores in considering his experience of life and death in the Nazi concentration camps of Theresienstadt and Auschwitz [191].

The 'delusion of reprieve,' Frankl notes, drawing on his earlier psychiatric experience, is one common response to an impending, apparently inevitable threat to our existence. Surely there has been some terrible mistake. Surely the cavalry will soon arrive with a rescue plan or pardon. And surely, to update the analogy, some technological wizardry will save us from crashing over the edge of the climate change cliff?

Frankl points to apathy, cynicism and despair; 'the blunting of emotions' and the 'dimming of reality' as reactions even more perilous than 'the delusion of reprieve.' 'Those who know how close the connection is between the state of mind of a man — his courage and hope, or lack of them — and the state of immunity of his body will understand that the sudden loss of hope and courage can have a deadly effect' [191, p. 84]. Anti-Nazi dissident and Lutheran Pastor Dietrich Boenhoeffer builds on Frankl's warning about the corrosive impact of cynicism in the following way.

> We have been silent witnesses of evil deeds; we have been drenched by many storms; we have learnt the arts of equivocation and pretence....intolerable conflicts have worn us down and even made us cynical.....What we shall need is not geniuses, or cynics, or misanthropes, or clever tacticians, but plain, honest, and straightforward men. Will our inward power of resistance be strong enough, and our honesty with ourselves remorseless enough, for us to find our way back to simplicity and straightforwardness? [192, pp. 16–17]

While US Congresswoman Alexandria Ocasio-Cortez might wish to gently but firmly update the gender specificity of Boenhoeffer's language she would probably agree strongly with him about the ongoing importance of courageous and straightforward honesty. As she noted in her recent comments on priorities for climate activists, 'the biggest hurdle that our communities have is cynicism – saying it's a done deal, who cares; there's no point to voting. If we can get somebody to care, it's a huge victory for the movement and the causes we're trying to advance' [193]. I am struck too in revisiting these comments by the deeply corrosive, impact of Donald Trump's profoundly cynical assessment, 'It is what it is' in relation to the capacity of government and citizens to work together to confront the public health calamity of COVID-19.

Frankl's review of strategies for sustaining the qualities which he calls 'inward power' resonates strongly with my more recent conversations with climate scientists and activists seeking ways of sustaining strength and meaning in the harshest of circumstances. The inspiration we draw from creative work and from our encounters with beauty and nature. Laughter and friendship and the love of other human beings. And most importantly our capacity, even in the most hopeless of situations to rise above our fear by giving ourselves 'to a cause to serve or another person to love....' [191, p. 115].

> We who lived in concentration camps can remember the men who walked through the huts comforting others, giving away their last piece of bread. They may have been few in number, but they offer sufficient proof that everything can be taken from a man but one thing: the last of the human freedoms—to choose one's attitude in any given set of circumstances, to choose one's own way. And there were always choices to make..... [191, p. 75]

Noting the starkly different circumstances experienced by those of us living in the affluence and comfort of the fossil fuel economy, Frankl might also be puzzled as to why so many of the most fortunate individuals in the wealthiest societies in human history are so easily tempted by strategies for avoiding responsibility for the consequences of our actions. Why do so many of us choose apathy, cynicism and 'the delusion of reprieve' as mechanisms for avoiding the decisive actions required to step back from the climate change abyss?

The Responsibilities of Freedom

Existentialism, Jean Paul Satre noted has often been unfairly misunderstood as a doctrine of gloomy nihilistic pessimism. To the contrary Satre asserted, the emphasis in existentialism on human solidarity and companionship and on the capacity and responsibility of every human being to decide their own course of action makes this 'the most optimistic of philosophies.'

In *Existentialism is a Humanism,* Satre argues that the defining characteristic of human beings is that each individual has the freedom and the responsibility to decide what matters to us, to set our own goals and to take the actions required to achieve them [194][191]. While our decisions are always affected by our circumstances, we do always have real choices. The complexity of the choices we face and the awareness that we can never fully know the consequences of our decisions may cause us deep distress. We cannot, in the end however avoid the realisation that it is our individual undertakings, our decisions and our actions which define and give meaning to our lives.

Our choices as individual human beings, Satre argues, are also always made in the context of our awareness that no human being exists alone. 'I cannot obtain any truth whatsoever about myself, except through the mediation of another. The other is indispensable to my existence, and equally so to any knowledge I can have of myself' [194, p. 9]. This leads Satre to a conclusion similar to the one arrived at by Kant. In making difficult decisions, 'one ought always to ask oneself what would happen if everyone did as one is doing.'

Satre illustrates and tests this proposition through the story of a young man who has to decide, in the darkest days of the German occupation of France, between staying to look after his ailing mother or escaping over the border to Spain to join the Resistance. The young man confronts two starkly different ways of acting: 'the one concrete, immediate, but directed towards only one individual; and the other an action addressed to an end infinitely greater, a national collectivity, but for that very reason ambiguous – and it might be frustrated on the way' [194, p. 6]. While Satre fully recognises the complexity of these choices he argues that the crucial challenge facing the young man is that he makes a conscious and deliberate choice rather than that hoping that someone else will make this tough decision for him.

Understanding the grave consequences of failure to take climate action at sufficient speed also leads us to a wide array of difficult decisions. To what extent should we change our individual consumption of energy or meat or airline travel? In what ways are we prepared to change the way we vote; the taxes we pay; the companies in which we invest our money? How much time and energy are we willing to devote to political campaigning; to organising protest movements or to civil disobedience?

Or, to take an example from the larger public arena of climate change policy making, imagine you are a government Minister confronted with the request to authorise massive investment in trialling and implementing geoengineering technologies. You are now thoroughly convinced that current emission trends are leading rapidly to global temperature increases likely to trigger catastrophic climate change impacts and tipping points. You have attended too many indecisive cabinet meetings and international climate conferences to have any faith left in the potential for decisive climate action at the necessary speed or scale.

Your advisers confirm that we are very close to the point of no return. Massive investment in negative emissions and planetary cooling they argue is now necessary to maintain ecologies capable of sustaining human civilisation. You see their point. You understand the physics. But you are also deeply concerned about the unpredictable, perhaps even more catastrophic consequences of the irreversible geoengineering techniques being proposed. And then of course there is the moral hazard problem. The faster you open the lid of the geoengineering Pandora's box the more quickly you encourage the comforting delusion that a last-minute technofix will always appear to provide us with a last-minute reprieve. At four in the morning you stare at the ceiling, unable to sleep. The Cabinet Meeting is at nine.

Confronting the Violence of Othering

The work of feminist and existentialist author Simone De Beauvoir further deepens understanding of the strategies many of us employ in order to avoid responsibility for our decisions and actions [195]. Some of us, De Beauvoir suggests, never fully mature beyond the young child's reliance on parental advice for their moral compass. Some seek guidance from an external source of authoritative expertise—a priest, a guru, a political leader or perhaps some great philosopher. A third alternative is the nihilistic response in which 'conscious of being unable to be anything, we decide to be nothing' [195, p. 57]. The final two pathways identified by De Beauvoir are the pathways of the adventurer who values conquest and action for its own sake even if this means trampling on the freedom of others and the passionate man who is willing to treat all other human beings as objects and as means to an end in order to achieve his goal.

For De Beauvoir, the exercise of 'genuine freedom' means taking a position and choosing a side in ways which enable and strengthen the freedom of others as well as ourselves. This is not primarily an argument based on selfless altruism but on recognising that our lives and actions only gain meaning through reciprocal relationships with other human beings. Importantly for current discussions about the consequences of climate change for future generations these reciprocal relationships extend across space and time. This helps explain the motivation which many climate activists find in considering the implications of our current actions for our children and grandchildren.

De Beauvoir's reflections on the reciprocal relationship between our freedom and the freedom of others lead her to focus close attention on the ways in which powerful individuals (frequently men) oppress and dehumanise other human beings (frequently women) by treating them as sub-human objects. Canadian social activist and author Naomi Klein builds on De Beauvoir's insights about the dehumanising consequences of 'othering' in her 2016 Edward Said lecture, *Let Them Drown: The Violence of Othering in a Warming World* [196].

Klein begins by noting the many ways in which colonising governments and corporations have deployed strategies of 'othering' (defined by Said as 'disregarding, essentialising, denuding the humanity of another culture, people or geographical region') to rationalise and excuse the violent conquest of Indigenous communities, the occupation of their lands and the theft of their resources. Klein goes on to emphasise the ways in which strategies of dehumanisation and othering continue to underpin the preparedness of the wealthiest individuals in the wealthiest countries to wilfully ignore the suffering of refugees and the brutal consequences of climate change for the poorest and most vulnerable communities.

> Although climate change will ultimately be an existential threat to all of humanity, in the short term we know that it does discriminate, hitting the poor first and worst....A culture that places so little value on black and brown lives that it is willing to let human beings disappear beneath the waves, or set themselves on fire in detention centres, will also be willing to let the countries where black and brown people live disappear beneath the waves, or desiccate in the arid heat. [196]

Mending What Has Been Torn Apart

In *The Myth of Sisyphus* Albert Camus draws on the story of the ancient Greek king, Sisyphus to explore the question which he describes as 'the one truly serious philosophical problem' [197, p. 5]. How do we judge whether life is or is not worth living? How do we sustain meaning and act with wisdom and courage in the face of overwhelming evidence that in the end all of our passion and our effort will end in darkness?

Sisyphus, Camus recalls, attempted to defy the gods by chaining the God Thanatos (Death) so human beings did not need to die. The Gods punished Sisyphus by condemning him to endlessly push a great rock to the top of a mountain. As soon as he reaches the top of the mountain, the rock immediately returns to the base of the mountain where Sisyphus must go to start his work over and over again.

How and why, Camus wonders, does Sisyphus find the strength to continue his task with no apparent end or purpose? Is it possible to imagine Sisyphus finding some level of happiness as he trudges back down the mountain to renew his labours? How do we as individual human beings find the courage to keep climbing and descending; breathing and acting in circumstances made even more absurd by our knowledge that all our work will end in death?

Faith in God and the promise of eternal life in heaven provides an answer for some. But for those of us without such faith what sources of wisdom and comfort can we draw on? Camus weighs the strengths and limitations of various options. Romantic love and hedonistic passion. The intense engagement of theatrical performance. The thrilling conquests of the adventurer or soldier. And the creative joyfulness of the artist.

While all of these possibilities have their advocates and compensations they are, in Camus's view all, in the end elaborate ways of distracting ourselves from the stark reality of our predicament. The only honest, authentic response lies in rejecting false hope and escapism and in constantly rebelling against the absurdity of the task we have been given.

Our task as [humans] is to find the few principles that will calm the infinite anguish of free souls. We must mend what has been torn apart, make justice imaginable again in a world so obviously unjust, give happiness a meaning once more to peoples poisoned by the misery of the century. Naturally, it is a superhuman task. But superhuman is the term for tasks [we] take a long time to accomplish, that's all. [198, p. 135]

In Camus's 1947 novel *The Plague* the characters in the plague ravaged town of Oran struggle to find motivation and direction for their actions despite their awareness that the epidemic overwhelming the town is far beyond their control [199]. The priest relies on his faith in God. The writer employs his skill at story-telling to bear witness to pain and despair. The teacher maintains his commitment to building knowledge and understanding. The physician, Dr. Rieux focuses on reducing suffering. Dr. Rieux knows that while he might save a few lives and minimise pain for a short time he cannot ultimately triumph over the remorseless spread of the disease. That is not in his view however any reason to give up. 'One must fight, in one way or another, and not go down on one's knees.'

As Australian social theorist Clive Hamilton notes in his 2011 book, *Requiem for a Species: Why We Resist the Truth About Climate Change* Dr. Rieux's commitment to 'active fatalism' and 'pessimism as strength'; of refusing 'to capitulate to hopeless odds, moving forward in the dark, feeling one's way and trying to do good' resonate strongly with many current conversations about meaning and action in a burning world [200]. Environmental educator Alison Keimowitz's meditation on her experience of being diagnosed with leukemia alongside her struggle to deal with the species wide existential threat of climate change explores similar terrain.

Returning to her teaching work on climate change and ecological risks after her recovery from leukemia (as a result of the unexpected gift of a stem cell transplant) Keimowitz is struck by the ways in which the experience of facing her illness has provided her with the resources and the wisdom to face a harsh climate future with honesty, grace and courage.

The years I've gained, however few or many they may be, are precious beyond measure. So too with the Earth. Each generation of humans living in relative abundance, each species saved from extinction for another 50 years, and each wild place left to function and inspire in its wildness, is precious beyond measure.

Inevitably, the climate will warm; whole ecosystems will be lost; and someday, there will be a last generation of humans on Earth. But the years we can postpone each loss, and each wild place and creature saved, are incalculably valuable. And so I keep teaching, and processing, and working to stave off the inevitable. I don't know if any of those things will truly prevent catastrophic changes on Earth; I suspect not. But I give these gifts freely, hopefully, and in the knowledge that they are all I have to give. [201]

Join up, Protest, Propose, Create

In revisiting Camus' central question about the primary purpose of philosophy: 'deciding whether life is or is not worth living' we are increasingly faced with the equally confronting dilemma: to what extent should we resist the collapse of our current way of life if it is precisely this way of life which is driving our species and many others over the cliff? Many of the toughest issues arising from this question are provocatively explored in the following debate between *Guardian* columnist George Monbiot and Paul Kingsnorth, founder of the *Dark Mountain project,* 'a network of writers, artists, and thinkers who have stopped believing the stories our civilisation tells itself: the myth of progress, the myth of human separation from nature; the myth of civilisation' [202].

Kingsnorth begins this discussion by noting that, in his view 'no amount of ethical shopping or determined protesting' is going to stop the inevitable collapse of industrial civilisation. The priority now, he argues, is not to offer false hope that we can 'shore up a crumbling empire with wave machines and global summits but to start thinking about how we are going to live through its fall, and what we can learn from its collapse' [202].

While generally accepting Kingsnorth's diagnosis of the causes and dangers of impending climatic and ecological disaster, Monbiot is fiercely opposed to the idea of 'accepting and perhaps even welcoming the likely collapse of industrial civilisation as some kind of cleansing fire.' How, he asks can we look away from the stark reality that 'while some human beings may survive the impending collapse the consequences for other species are likely to be much worse…When civilisations collapse, psychopaths take over….' [202]

'This is why,' Monbiot writes 'despite everything, I fight on. I am not fighting to sustain economic growth. I am fighting to prevent both initial collapse and the repeated catastrophe which follows from it. However faint the hopes of engineering a soft landing – an ordered and structured downsizing of the global economy – might be, we must keep this possibility alive' [202]

Kingsnorth replies that Monbiot has too simplistic a view in imagining only two potential futures: Liberal Capitalist Democracy 2.0 (with a democratically governed steady-state economy fuelled by renewable energy) or a terrifying cannibalistic 'McCarthy World' based on Cormac McCarthy's post-apocalyptic novel, *The Road*. Surely, Kingsnorth suggests, we can envisage and create saner and more desirable alternatives to industrial civilisation than this.

Monbiot's reply focuses attention on the question of how many people the world could support without either fossil fuels or an equivalent investment in alternative energy? His deep concern (no doubt informed by John Schellnhuber's chilling calculations about the population which could be supported in a world in which global temperature had soared beyond four degrees) is that that billions of human beings would suffer and perish in any rapid civilisational collapse. Monbiot accepts that the chances of a successful climate emergency rebellion are small. But, as he rightly notes 'they are non-existent if we give up before we have started.' His call to action is clear and strong. 'Join up, protest, propose, create. It's messy, endless and uncertain of success….but it's all we've got and all we've ever had….' [202].

Messy, endless and uncertain of success indeed. Critique and solidarity; freedom and responsibility; resistance and rebellion. Scaling the mountain as the storm clouds swirl around us, building the staircase as we climb. Guided as best we can by history and ethics; art and science; strengthened and supported by the care and wisdom of our fellow climbers. Some of course would argue that there are indeed other pathways up the mountain, other sources of hope and courage, guided by faith and prayer and spirituality as well as science, reason and justice.

7
Mercy to All Beings: Learning from Christian, Jewish and Islamic Traditions About Thankfulness, Love and Care

In June 2019, over 150 Australian religious leaders from many faith traditions wrote an open letter to the newly elected Australian Prime Minister, Scott Morrison urging him to make the climate emergency his number one priority. Morrison, a long-standing member of the Pentecostal Church, is also infamous in Australia for parading a lump of coal into parliament in order to demonstrate his passionate support for the coal industry.

Noting the urgency of ending coal mining as an essential basis for decisive climate action, the Bishops, Ministers, Rabbis and Muftis called on Morrison to support the three key demands of Australia's school climate strikers: stopping the construction of the huge Adani coal mine; committing to no new coal or gas projects in Australia; and moving to 100% renewable energy by the year 2030. The religious leaders were no doubt disappointed but unsurprised by the Morrison government's failure to change direction on any of these issues.

Some devout believers may find the role of religious faith in sustaining hope and courage; meaning and purpose self-evident. Non-believers and atheists often take a very different view, noting the ways in which some politicians and business leaders point to religious ideas and beliefs as a convenient excuse for avoiding and denying the urgency of climate action.

I come to these conversations as a person long departed from my Protestant upbringing and well aware of the criticisms which many non-believers have of the ways in which religious faith has, at times, been used to justify violence and persecution. I retain considerable respect however for the ethical and ecological wisdom of writers and activists from a variety of religious and spiritual traditions. In doing so, I share Pope Francis' view that the complexity of the ecological crisis calls for recognition that 'solutions will not emerge from just one way of interpreting and transforming reality. Respect must also be shown for the various cultural riches of different peoples, their art and poetry, their interior life and spirituality' [203, p. 45].

This chapter therefore explores ideas and insights on meaning and courage in challenging times by exploring responses to the climate crisis which resonate most strongly for me from Christian, Jewish and Islamic writers and activists. The central question underpinning this discussion, the ongoing debate between human-centric and ecological religious perspectives and paradigms, is concisely summarised in the following observations from Mary Evelyn Tucker and John Grim, co-founders of the *Yale University Forum on Religion and Ecology*.

> The question arises whether the wisdom traditions of the human community, embedded in institutional religions and beyond, can embrace integral ecology at the level needed? Can the religions provide leadership into a synergistic era of human–Earth relations characterized by empathy, regeneration, and resilience? Or are religions themselves the wellspring of those exclusivist perspectives in which human societies disconnect themselves from other groups and from the natural world? Are religions caught in their own meditative promises of transcendent peace and redemptive bliss in paradisal abandon? Or does their drive for exclusive salvation or truth claims cause them to try to overcome or convert the Other? [204, p. 6].

The Cry of the Earth and the Cry of the Poor: Christianity, Ecology and Climate Change

Many current debates about the role of Judeo-Christian religious traditions in a time of climate crisis begin by referencing the continuing relevance of American historian Lynn White's article, *The Historical Roots of Our Ecologic Crisis* first published in the journal *Science* in 1967 [205]. Writing in the years immediately following the publication of Rachel Carson's *Silent Spring*, White argued that Biblical teachings about the mastery of human beings over nature, articulated most clearly in the first pages of *Genesis* (1: 27–28), have been powerful, foundational drivers of accelerating ecological risks and crises.

> God created humankind in his image, in the image of God he created them; male and female he created them. God blessed them, and God said to them, 'Be fruitful and multiply, and fill the Earth and subdue it; and have dominion over the fish of the sea and over the birds of the air and over every living thing that moves upon the Earth.'

White argues that the widespread influence of these Biblical ideas about human supremacy swept away more ancient animistic recognition and respect for the spirits inhabiting 'every tree, every spring, every stream and every hill.....Man's effective monopoly on spirit in this world was confirmed, and the old inhibitions to the exploitation of nature crumbled' [205, p. 1205]. White goes on to suggest that an alternative and arguably more desirable religious and ecological paradigm can be found in the teachings of the thirteenth-century Italian friar Saint Francis of Assisi.

Saint Francis proposed a very different lens through which Christians could view the relation between human beings and nature: 'the substitution of the idea of the equality of all creatures, including man, for the idea of man's limitless rule of creation' [205, pp. 1206–1207]. 'The key to understanding the ideas of Saint Francis is his belief in the virtue of humility, not merely for the individual but for man as a species. Saint Francis' view of nature and of man rested on the idea that all things animate and inanimate are designed for the glorification of their transcendent Creator' [205, pp. 1206–1207]. White would no doubt have been intrigued by Argentinian Jesuit Cardinal Jorge Mario Bergoglio's decision to take the name of Francis in honour of Saint Francis of Assisi when he became Pope in 2013 and by the publication, two years later of the Papal Encyclical on climate change and ecology, *Laudato Si: On Care for Our Common Home*.

'I believe' Pope Francis writes, 'that Saint Francis is the example par-excellence of care for the vulnerable and of an integral ecology lived out joyfully and authentically. He is the patron saint of all who study and work in the area of ecology, and he is also much loved by non-Christians. He was particularly concerned for God's creation and for the poor and outcast. He loved, and was deeply loved for his joy, his generous self-giving, his openheartedness' [203, p. 9]. While not generally inclined to leaf through Papal Encyclicals, I have, like many other secular readers, found *Laudato Si* to be a valuable source of wisdom about the importance and urgency of swift and decisive climate action.

The title of the Encyclical, *Laudato Si* ('Praise be to you') is inspired by the opening words of Saint Francis' famous prayer, *The Canticle of Brother Sun and Sister Moon* [206]. This lyrical poem might also usefully be forwarded to the world's political and economic decision makers as an anthem to the beauty of the world we are seeking to celebrate and protect.

> Praised be You my Lord with all your creatures, especially Sir Brother Sun, who is the day through whom you give us light. And he is beautiful and radiant with great splendour, of you most high, he bears the likeness.
> Praised be You, my Lord, through Sister Moon and the stars, In the heavens you have made them bright, precious and fair.
> Praised be You, my Lord, through Brothers Wind and Air, and fair and stormy, all weather's moods, by which You cherish all that You have made.
> Praised be You my Lord through Sister Water, so useful, humble, precious and pure.
> Praised be You my Lord through Brother Fire, through whom You light the night and he is beautiful and playful and robust and strong.
> Praised be You my Lord through our Sister, Mother Earth who sustains and governs us, producing varied fruits with coloured flowers and herbs.

Laudato Si begins by acknowledging the profoundly distressing evidence of the many ways in which 'the Earth, our sister is crying out to us because of the harm we have inflicted on her, by our irresponsible use and abuse of the goods with which God has endowed her. We have come to see ourselves as lords and masters [of the Earth] entitled to plunder her at will' [203, p. 3]. The Encyclical explicitly rejects simplistic interpretations of the proposition that the belief that human beings have been created in God's image justifies the claim that we have any right or obligation to dominion over other creatures.

> The biblical texts are to be read in their context, recognizing that they tell us to 'till and keep' the garden of the world. 'Tilling' refers to cultivating, ploughing or working, while 'keeping' means caring, protecting, overseeing and preserving. This implies a relationship of mutual responsibility between human beings and nature. [203, p. 49]

For Pope Francis, affirmation of an ethic of stewardship and of interdependence between human beings and nature requires a carefully balanced understanding that a 'misguided anthropocentrism need not necessarily yield to biocentrism….Human beings cannot be expected to feel responsibility for the world unless, at the same time, their unique capacities of knowledge, will, freedom and responsibility are recognized and valued' [203, p. 88].

Laudato Si also argues that it is both unrealistic and immoral to expect the poorest and most brutally exploited individuals and communities to support decisive climate action without addressing the underlying causes of injustice and inequality.

> The climate is a common good, belonging to all and meant for all. It is therefore immoral to act in such a way as to generate changes in the climate that affect others— especially those who because of poverty cannot adjust or adapt….A truly ecological approach must integrate questions of justice in debates on the environment, so as to hear both the cry of the earth and the cry of the poor. [203, p. 35]

Just and effective responses to the complex ecological and climate crises now unfolding around us therefore depend on our capacity to move beyond technocratic and consumerist, cultures and economies. Key principles of the 'integral ecology' guiding this transformative practice include recognition that care for creation is a virtue in its own right and the necessity of a new ethic of global solidarity guiding and directing our search for the common good.

The Encyclical concludes with a series of reflections on ideas and values which may help us traverse the difficult terrain that now lies before us: a spirit of generous care and thankfulness for the gifts of life and the beauty of our common home; heightened awareness of our common origins, our mutual belonging and our interdependence with other creatures; a simpler, more contemplative lifestyle, capable of deep enjoyment and free of obsession with consumption; and renewed understanding of the importance of love, compassion and respect in social, political, economic and cultural life.

We are speaking' Pope Francis reflects 'of an attitude of the heart, one which approaches life with serene attentiveness, which is capable of being fully present to someone without thinking of what comes next, which accepts each moment as a gift from God to be lived to the full' [203, p. 165]. Even as a non-believer I find these reflections, which also resonate strongly with Buddhist and Taoist ideas and principles, relevant and useful.

The Smoke of Grief and the Fire of Love

Jim Antal is a Church of Christ Minister, climate activist and theologian who serves as a special adviser on climate justice to the President of the *United Church of Christ* in the United States. In 2019, he received the Steward of Creation Award from the *National Religious Coalition for Creation Care*. Previous recipients of this award include climate scientist James Hansen, ecological philosopher Wendell Berry, climate activist Bill McKibben and the Sioux Nation. In reviewing Antal's book *Climate Church Climate World*, Archbishop Desmond Tutu writes, 'Jim Antal shows how the church can engage the urgent moral crisis of climate change. This book will inspire both the courage and conviction people of faith need to provide the leadership necessary to realise God's dream of a just world in which humanity is reconciled to all creation' [207].

Climate Church Climate World draws on the same commitment to care for all creation celebrated in *Laudato Si*. Life on Earth, God's great gift of creation faces grave and accelerating risks as the direct result of human hubris and greed. 'Those who follow Jesus' Antal argues 'will not back away from God's call to protect our common home.' Gratitude for the gift of creation is, in Antal's view, an enormously powerful motivating force for climate action. 'Gratitude for having been given life; gratitude for God's creation and all the ways it has nourished one's life; gratitude for the support that friends and loved ones have provided; gratitude for this particular moment, as well as the gift of time itself; gratitude for the dreams and aspirations that mysteriously arise from within' [207].

The tragic dilemma we now face is that awareness of the beauty of the world too often deepens our grief and heightens our fear in confronting the harsh realities of mass extinctions and climate crisis. Grief and fear can be powerful catalysts for action so long as they do not cross over the line into despair, paralysis and panic.

> To become people of hope we must be willing to stare reality in the face. We must be willing to face not only the scientific reality of a rapidly warming world but also the political reality that there are individuals, groups, and entire industries devoted to spreading misinformation and lies about the climate crisis. The smoke of grief and the fire of love are inseparable. The more we enlarge our hearts with love and gratitude for the gift of God's life-giving and life-preserving creation, the more our hearts will fill with the courage we need to overcome our fear. [207, p. 157]

Antal proposes four theological cornerstones which might underpin a Christian framework for climate justice action. First, that the covenant between human beings and God is everlasting for all creatures and for all time. Second that the Golden Rule, the idea that we called to love our neighbours as ourselves should be extended to include all future generations and all other species. In thinking about the scope of our climate justice responsibilities, we should therefore keep in mind the words of Dr. Martin Luther King: 'we are caught in an inescapable network of mutuality, tied in a single garment of destiny. Whatever affects one directly, affects all indirectly.' Third, that honouring creation requires us to fulfil the responsibilities of ecological stewardship and intergenerational obligation with the open-hearted generosity and compassion exemplified in the life and teachings of Jesus. Fourth, that understanding and recognising the interdependence of human beings with the rest of creation mean that our responsibility for collective action is at least as important as the individual salvation.

These theological cornerstones Antal suggests can help us imagine and enact ways of life and political practice better aligned with the scope and scale of the climate emergency and ecological challenges we now face. These principles might, for example, lead us to explore and build social and economic relationships and institutions informed by values of 'collaboration in place of consumption; wisdom in place of progress; moderation in place of excess; vision in place of convenience; accountability in place of disregard and self-giving love in place of self-centered fear' [p. 81].

They might also inspire and embolden consideration of actions such as divestment and non-violent civil disobedience which directly challenge the social license of vested interests standing in the way of accelerating the transition to a just and resilient zero-carbon future. As Antal notes in his concluding reflections 'when considering civil disobedience and the other forms of witness, people have shared with me that love is their most powerful motivator — love of God; love of nature; love of beauty; love of their children; love of creatures and plants in all their diversity; love of the impossible way in which this planet provides all living things with everything we need to flourish' [207, p. 144].

Faith and Evidence; Suffering and Care

Dr. Katharine Hayhoe is director of the *Climate Science Centre at Texas Tech University* and author of over 120 peer reviewed climate science articles as well as numerous reports for the US National Academy of Sciences and the Intergovernmental Panel on Climate Change. In 2014, she was chosen as one of *Time Magazine's* 100 most influential people and received the American Geophysical Union Climate Communication award. Dr. Hayhoe is also a devout evangelical Christian and coauthor with her evangelical pastor husband Andrew Farley of *A Climate for Change: Global Warming Facts for Faith-Based Decisions* [208].

Hayhoe has, not surprisingly, come under savage attack from some evangelical Christians fiercely critical of climate science. She continues to speak however with calm and confident authority about the ways in which her Christian faith is not only consistent with her commitment to climate science but also helps her sustain resilience and courage in the face of mounting evidence of climate risk.

When asked by climate sceptics whether she believes in global warming, Hayhoe begins by saying, 'No, I don't.....because belief is faith; faith is the evidence of things not seen. Science is evidence of things seen. To have an open mind, we have to use the brains that God gave us to look at the science.' 'A thermometer,' she often notes, 'is not Liberal or Conservative.'

> I'm a climate scientist. When I'm asked about global warming, my answer is unequivocal: It's real, we're causing it, and it's serious. Every week, I receive bile-filled messages. They accuse me of getting rich off my research, or perpetuating a hoax, or even aiding the Antichrist. I get these messages because I'm stating the truth about what's happening to our planet.
>
> I'm not committed to this only because I'm a scientist, but because I'm a human. I'm a mother who wants a safe world for her child to grow up in – and everyone else's as well. I'm a lifelong Christian who believes that we should love others as Christ loved us and care for those who are suffering, their vulnerability exacerbated by a changing climate. [209]

Hayhoe's commitment to decisive climate action and to climate justice is informed and inspired by her Christian faith. 'If you believe that God created the world, and basically gave it to humans as this incredible gift to live on, then why would you treat it like garbage? Treating the world like garbage says a lot about how you think about the person who you believe created the Earth.'

Hayhoe often begins her public presentations by quoting John Holdren, President Obama's science adviser: 'We basically have three choices: mitigation, adaptation, or suffering. We're going to do some of each. The question is what the mix is going to be. The more mitigation we do, the less adaptation will be required and the less suffering there will be' [210]. 'As scientists,' Hayhoe argues 'we don't know a lot about suffering, but as Christians we do.'

> We know that part of the reason we're here in this world is to help people who are suffering. And that suffering will not be meted out proportionally: if global warming continues unchecked, the poor – whether they're in Houston's Fifth Ward or in low-lying areas of Bangladesh – who have contributed least to carbon emissions will feel the most pain, from enduring more intense heat waves to paying the higher food prices that will accompany failed crops.' [210]

Throughout the Bible, Hayhoe reminds her audience, Christ's teachings emphasise the power of love and the importance of service to others: 'In the same way I loved you, you love one another. This is how everyone will recognize that you are my disciples—when they see the love you have for each other.' (John 13:34–35)

Hayhoe's commitment to honestly facing and communicating the implications of scientific evidence leaves her in no doubt about the speed and scale of impending climate risks. In her more optimistic moments, she envisages a gradual bend away from the emissions trends leading towards the most catastrophic climate scenarios until the point at which worsening climate disasters eventually lead to a collective 'oh shit!' moment. Perhaps, she reflects, this could be the point at which the world suddenly ramps up its climate action to the scale of a Manhattan Project or a moon shot [211].

Hayhoe places her greatest hope for the decisive actions required to bend emissions downward on providing a vision of the positive benefits of a rapid transition from fossil fuels to a clean energy economy.

> We need to hear the stories of real people, making real-life choices today, that save money, improve our health, increase our energy independence, grow local jobs, and help those less fortunate than us. So I'm a huge fan of solutions like reducing food waste, which also tackles hunger; smart grazing strategies that increase soil carbon uptake and restore

degraded lands to support people's livelihoods; and the expansion of new clean energy in areas where people don't have cheap and easy access to electricity. [209]

Between the Fires: Judaism, Ecology and Climate Change

Rabbi Authur Waskow, founder of the *Shalom Centre for Conflict Resolution* and Reconciliation, has been writing, speaking and campaigning on social justice, civil rights, peace and ecological issues for over fifty years. A prolific author and highly respected teacher, he also wears each of his 26 arrests for civil disobedience as a badge of honour. In doing so, he cites Rabbi Abraham Joshua Heschel's reflection following his participation in the 1965 Selma Civil Rights march: 'I felt as if my legs were praying.' Rabbi Waskow has been named by *Jewish Daily Forward* newspaper as one of 'America's most inspiring rabbis.' He has also been awarded the Peace and Justice Award by the Muslim American Society Freedom Foundation.

Informed by his increasingly deep concern about the implications and responsibilities of the climate emergency, Rabbi Washkow has written the following *Prayer for Lighting Candles of Commitment*. The Prayer draws on and is inspired by traditional Jewish 'midrash' (commentaries on the Hebrew scriptures) about the dangers of Flood or Fire. As a reader not schooled in Jewish scriptures or history, I find the poetry of this prayer a strong and moving illustration of the work which writers from many traditions are undertaking to make sense of a rapidly changing, increasingly threatening world.

Between the Fires begins by acknowledging and confronting the realisation that 'we are the generation that stands between the fires.' Behind us lie 'the flame and smoke that rose from Auschwitz and from Hiroshima.' All around us now we see the consequences of burning forests, melted ice fields, flooded cities and scorching droughts. 'Before us lies the nightmare of a Flood of Fire, the heat and smoke that could consume all Earth....' Washkow's prayer continues by providing the following suggestions about ways of living in these times between the fires.

It is our task to make from fire not an all-consuming blaze but the light in which we see each other fully. All of us different, All of us bearing One Spark.
We light these fires to see more clearly that the Earth and all who live as part of it are not for burning.
We light these fires to see more clearly the rainbow in our many-colored faces. Blessed is the One within the many. Blessed are the many who make One. [213]

Dr. Hava Tirosh-Samuelson, Director of Jewish Studies at *Arizona State University* and author of *Judaism and Ecology: Created World and Revealed World*, argues that to speak authentically from the sources of Judaism, one must affirm that God created the world [and that] In the created order, the human being is given a privileged place. It is however, he continues, citing the following verses from *Ecclesiastes*, precisely because humans are created with the capacity to transcend nature that they are commanded by God to protect nature' [214, p. 102].

When God created Adam, God led him around all of the trees in the Garden of Eden. God told him, 'See how beautiful and praiseworthy are all of my works. Everything I have created has been created for your sake. Think of this and do not corrupt the world; for if you corrupt it, there will be no one to set it right after you.' (*Ecclesiastes* Rabbah 7:13)

Tirosh-Samuelson outlines the implications and relevance of core Judaic principles for informing and prioritising ecological and climate action in the following way. The principles of *Bal tashchit* (do not destroy) and *Tza'ar ba'aley hayim* (minimise distress to living creatures) outlined in *Deuteronomy* provide a stern warning against acting in ways which lead to the destruction of fruit trees; overgrazing pastures; unjustifiably killing animals; hunting animals for sport; polluting air and water; the extinction of species; and the overconsumption and squandering of natural resources. The principles of *Tikkun Olam* (healing the world) and *Hilkhot Shekeinim* (the law of neighbours) can also help illuminate the importance of limiting and repairing the damage caused to human beings and to other species by hubris, selfishness and greed.

The rituals of *Shabbat* (the Sabbath day of rest and prayer) and of *Shmitta* (the practice of allowing the land to lie fallow once every seven years) can foreground the need to balance short-term gains resulting from exploiting the resources of the natural world with the longer-term benefits of protecting the welfare and wellbeing of human and non-human species. A number of Jewish communities have also broadened the scope of *kashrut* (the rules governing what is fit to eat) to encompass the practice of *Eco-kashrut* in order to help pay closer attention to the ecological and social justice impacts of the ways in which food is produced and consumed.

There is also, as Rabbi Rami Shapiro notes in his widely cited interpretation of the *Pirke Avot (The Sayings of the Fathers)*, considerable wisdom and comfort to be found in learning the art of combining passionate commitment to climate justice with awareness of the limitations of the outcomes which each of us acting alone can achieve. 'Do not be daunted by the enormity of the world's grief. Do justly, now. Love mercy, now. Walk humbly, now. You are not obligated to complete the work, but neither are you free to abandon it' [215, p. 41].

We Will Again Stand Mountain-Strong: *Tisha B'Av* in a Time of Climate Crisis

Tisha B'Av, one of the most sorrowful days in the Jewish religious calendar, commemorates the date on which the first and second Jewish temples in Jerusalem were destroyed, first by the Babylonians and then the Romans. In more recent times, the day provides an opportunity to reflect on other experiences of loss and suffering experienced by the Jewish people. *Tisha B'Av* is traditionally observed as a day of fasting along with the public reading of lamentation poems known as kinot.

In 2010, the *Shalom Centre* commissioned as a kinot a *Lament for the Earth* enabling *Tisha B'Av to* also be observed as a day of mourning for 'the danger now faced by our universal Temple Earth.' While the impetus for this work was the 2010 Gulf of Mexico oil spill from the BP oil rig Deepwater Horizon, the Shalom Centre *Tisha B'Av* commemoration also focuses attention on suffering arising from the climate crisis and the urgency of action to halt the damage caused by modern-day fossil-fuel corporate empires.

The *Lament* opens with a cry of anguish and a call to action. 'As we stand on the brink of the burning of not just the Temple, but also our world....help us to see that we are not alone, powerless against a global problem. May we realize that our cries, though they come from the depths, will be heard. Our lives, rewoven together, can make a difference' [216]. *The Lament* continues by acknowledging that 'when tragedy strikes, or even when a crisis looms, It is natural to look the other way, to deny, to disbelieve, to pretend that life can go on as before....Denial is natural, but futile, even dangerous. The carbon curve climbs, the waters rise, the fires rage – there is such a thing as too late.'

> To this personal denial, those who profit from the status quo add deception. They lull us with the poppy-milk of false prophecies. It's not happening; humans aren't causing it; and any way it won't be so bad. Or if it is, we can trust in technology to find a painless solution. With deception and delusion, we are distracted from justice. But justice delayed and justice denied bring the sword into the world. This day, we recommit to the pursuit of justice. [216]

The third Section *Overwhelming Depression* asks us to recall that 'we are far from the first age to ask why suffering continues and why God has forsaken us. Yet it seems that now that depression, grief and anxiety have become increasingly pervasive, leading to the question, what about our age is uniquely overwhelming? Is it what we face externally, a climate spinning out of control, forests burned, species driven to extinction? Perhaps it is what is between us, deepening divisions and splintering societies, or an internal void, a loss of meaning coupled to our loss of connection.'

In the fourth section of *The Lament, Blame and Responsibility,* the authors wonder whether *Tisha B'Av* can be an important time for soul-searching and for deepening our understanding of individual and collective responsibility. 'Even against the flooding of coasts and the loss of species, we cannot take refuge in blame or powerlessness. It is time to accept our call, to shoulder responsibility to undertake the healing of the world' [216]. *The Lament* concludes with the reflection that 'mourning is different from despair, from giving up. To the contrary, only when we fully understand how broken our hearts are can we begin to find a path toward healing.'

> If, with clear eye and strong heart, we face disaster unflinching,
> Strengthened by each other, inspired by the Breath of Life
> It may yet come to pass that we will again stand mountain-strong
> Our dirge turned into dance, sackcloth undone and bound instead with joy (Psalm 30:12)

The World Is Sweet and Verdant: Islam, Ecology and Climate Change

While I am even less familiar with Islamic teaching than I am with Christian and Jewish texts, my initial encounters with ideas which Islamic scholars draw on from the *Qur'an* to inform and guide Islamic climate change and ecological action suggest some strikingly familiar themes [217, 218].

The idea of *Tawhid* refers, for example, to the core Islamic doctrine of the oneness of God as creator and sustainer of the universe. In describing the heavens and the Earth as a divine symphony and a hymn of praise to God, the *Qur'an* (2:255) envisages all of creation, including human beings as one vast interwoven pattern of complex, interdependent elements.

Fitra describes the harmonious origins of creation and of humankind. In this view, the underlying patterns of creation cannot be altered. Overreaching our power in attempting to radically change these patterns may lead human beings on a pathway to self-destruction. According to *Mizan*, the principle of balance and the middle path, all creation has an order and all elements of creation have a special purpose. Human beings, the *Qur'an* tells us, have been created by God with special capabilities including the capacity to reason and to understand. These endowments give humankind the unique and sacred responsibility of celebrating and protecting the 'bounties of the Lord.'

The principle of *Khalifa* refers to stewardship of the Earth as the sacred duty God requires of the human species. Human beings in this view are guardians not masters of the Earth. This responsibility imposes strict limits on human ambition, requiring us to constantly remember and respect the complexity and fragility of our relationship with the rest of creation.

The Qur'an requires human beings to 'walk not exultantly upon the Earth' (17:63) and to view the whole of nature as 'signs for a people who hear (10:67), signs for a people who reflect (13:3; 30:21) and signs for a people who understand' (2:164). If we lose our capacity to see clearly and to listen with care, we run grave risks of 'working corruption upon the Earth' (*Qur'an* 2:205). 'Corruption has appeared in both land and sea because of what people's own hands have brought. So that they may taste something of what they have done, so that hopefully they will turn back' (*Qur'an* 30: 41). These teachings lead some ecologically minded Islamic scholars to the view that our rapacious exploitation of the Earth means we have indeed become like 'they who have hearts with which they understand not; they who have eyes with which they see not; and they who have ears with which they hear not' (7:179).

Professor Al-Jayyousi, Head of Sustainable Innovation at the *Arabian Gulf University* in Bahrain, suggests that the essence of Islam in fact lies in 'being in a state of harmony with the natural state *(fitra)* and in respecting balance *(mizan)* and proportion *(mikdar)* in the systems of the universe' [218]. In this view, every living species is a community *(ummah)* and is part of the community of life *(ummam)*. 'Every living thing is in a state of worship. When one hurts a bird or a plant, he or she is silencing a community of worshippers' [218]. The harmony and balance of God's creation have been disturbed due to human choices resulting in overconsumption, overexploitation and overuse of resources. 'Environmental problems, such as the destruction of natural habitats, loss of biodiversity, climate change, and erosion of soil are all triggered by human greed and ignorance. Human responsibility is to save and protect livelihood and ecosystem services to ensure a sustainable civilization learning from and reflecting on the fate of past civilizations' [218].

Prof Al-Jayyousi proposes an Islamic practice of Earth stewardship through an integrated practice of 'green activism, innovation and lifestyle.' An Islamic practice of ecological and climate emergency action can, he suggests, be understood as a struggle against the imbalances and injustices that disturb the 'natural state' (fitra). The practice of 'green innovation and green lifestyle' should reflect and be informed by Islamic teachings about the importance of living lightly on Earth (zohd); of limiting waste and extravagance (israf); of justice (adl); and of beauty (Ihsan).

Islam and Climate Change: A Call to Heal published in 2010 by the London based Islamic ecological forum *Wisdom in Nature* draws on the principles of ecological awareness, balance, responsibility and Earth stewardship outlined above to provide a concise overview of the Islamic justification for courageous and decisive climate action. 'In Islam is a powerful and timely message. If we listen to it, we will become aware of the consistency with which it teaches us to revere nature and to be sensitive to God's works in the form of animals, insects, the soil and its organisms, the clouds, the sun and moon. It teaches us to live in harmony with the natural order, to take only what we need and not to waste' [217].

The work of *Wisdom in Nature* is informed by a *Five Strand Activism Model*, for creating 'conscious, just and regenerative communities.' The Model aims 'to facilitate a movement away from states, processes, and paradigms that contribute to degradation of the social and wider ecology moving towards ones that are nurturing, wholesome, in alignment with the natural order (fitrah), and that help restore ecological balance (mizan).' Many of the *Five Strand Model's* key principles, summarised in the following way resonate strongly with ecological and climate action frameworks informed by a wide range of other spiritual and secular perspectives.

> *Earth and Community:* From corporate domination and consumerism towards simplicity, sharing and a deeper connection to the Earth and its diverse communities.

> *Deep Democracy:* From concentration of power amongst the rich and privileged towards equalisation of power that honours diversity, draws out consensus and creativity, and empowers all.
> *Whole Economics:* From monetary systems disconnected from real value and embedded in usury towards just economic systems nurturing to life, soul and community.
> *Climate Justice:* From dependence on fossil fuels towards non-polluting energy, needs above profit and socially just solutions.
> *Engaged Surrender:* A nonviolent, process-oriented activism, expressed through a contemplative dimension within the framework of Islam and in surrender to the underlying unity or oneness. [217]

The Islamic Declaration on Global Climate Change

The *Islamic Declaration on Global Climate Change* launched in 2015 and endorsed by a wide range of leading Islamic clerics, academics and policy makers begins with a hymn of praise strikingly reminiscent of the prayer of St Francis. 'The stars, the sun and moon and this Earth in all the diversity, richness, and vitality of its communities of living beings reflect and manifest the boundless glory and mercy of their Creator' [219].

> Excessive pollution from fossil fuels threatens to destroy the gifts bestowed on us by God, whom we know as Allah—gifts such as a functioning climate, healthy air to breathe, regular seasons, and living oceans. But our attitude to these gifts has been short-sighted, and we have abused them. What will future generations say of us, who leave them a degraded planet as our legacy? How will we face our Lord and Creator?

The Declaration places strong emphasis on the responsibilities of wealthy nations and oil-producing states in accelerating the phase out of greenhouse gas emissions and reducing consumption 'so that the poor may benefit from what is left of the Earth's non-renewable resources....realising that to chase after unlimited economic growth in a planet that is finite and already overloaded is not viable.'

The Declaration concludes by asking readers to bear in mind the words of the Prophet Muhammed. The world is sweet and verdant, and verily Allah has made you stewards in it, and He sees how you acquit yourselves.' Discussion of the implications of these ideas with an Islamic colleague also led him to share the wise advice first articulated by the Prophet Mohammed: 'If this is the last day of life on Earth and you have a small plant, make sure you plant it.'

Sometimes We Are Compelled to Act: Faith-Based Engagement with Extinction Rebellion

In April 2019, a young Muslim woman, Sarah Zaltash shared this prayer with *Extinction Rebellion* demonstrators gathered at London's Oxford Circus.

> Allahu 'akbar (God is Great) There is nothing greater than oneness....no H&M, no Piccadilly, none of that is greater than oneness...Hayya 'ala I-falah. Come to sanctuary....'Cause that's how important it is to stay safe...Hayya 'ala s-salah. Which means come celebrate. Come worship. Come pray. [221]

As she began to sing her prayer, joined by many in the crowd the police moved into begin arresting people. Zaltash halted her song and turned to the police asking them to stop, explaining to them that 'we are in the middle of prayer!' Concluding her prayer she invited the crowd 'to kiss the ground and place your forehead upon it three times.' 'Blessings to you all' she concluded. 'You are oneness.'

In October 2019, the UK-based organisation, XR Muslims published a statement outlining the Islamic case for joining *Extinction Rebellion* [222]. The statement begins with the recognition that 'our children and future generations are in jeopardy, due to the self-inflicted climate emergency.' It goes on to argue that, 'as Muslims we are responsible for building a society balanced with truth, equity, justice and compassion.

We have a duty to exercise our individual and a collective sacred responsibility to maintain the Quranic instruction of Al Mizan (balance) over Allah's creations.'

> To take action to restore health to our air, our soil, our water, is an act of remembrance of Allah *(*dhikr*)*, and an act of compassion towards Allah's creations *(*Rahmah*)*. Let us come together with Extinction Rebellion, with each other and people of other faiths, to bring *Al Mizan* (balance) and *Rahmah* (compassion) to our wounded planet, our wounded humanity. [222]

The April 2019 *Extinction Rebellion* occupation of Oxford Circus was also welcomed by Jewish demonstrators with 12 short blasts of the shofar, the Jewish religious musical instrument made from a lamb's horn. This action was intended to symbolise the 12 years climate scientists say humanity has left to prevent irreversible climate breakdown. Lianna Etkind, a leading member of *XR Jews,* spoke of the relevance to the demonstrators of the Jewish principle of l'*dor vador* in the following way. 'Jewish life is predicated on the idea of *l'*dor vador, of passing on our values and our stories to the next generation and Jewish continuity. And what happens when you can't rely on that continuity?' [223].

On the Friday night of the Oxford Occupation, 80 members of the *XR Jews* community came together to celebrate the Sabbath at the protest site in Parliament Square. They shared together the songs, food and drink of the Shabbat meal as symbols of the suffering Earth and as a celebration of solidarity and hope. Their Shabbat meal commemorating the Hebrews' exodus from slavery in Egypt included a version of the Passover song *Dayenu* ('it would have been enough') with an interpretation focusing attention on the importance of reducing consumption and lowering carbon emissions.

Rabbi Oliver Joseph read testimonies from First Nation and Global South communities already suffering climate change destruction. Rabbi Nathan Godleman and Rabbi Jonathan Wittenberg asked the gathering to remember that 'we are God's stewards, obligated to conserve Creation and to consider future generations. Sometimes we are compelled to act, to answer the call of the moment. Civil disobedience with a commitment to non-violence has prophetic roots' [223].

Members of *Christians for Climate Action* noted that the 2019 April XR Occupation was occurring during Holy Week, a time of year when Christians reflect closely on sacrifice, death and rebirth in the journey from despair to hope. They therefore organised a climate-themed *Stations of the Cross* walk between the road blocks at each end of the Marble Arch site occupied by XR [222]. The *Stations of the Cross* walkers concluded their final prayer by sitting in the middle of the road between the blockades and the area where the police vehicles were gathered. Asked to clear the road by a police officer, one activist politely asked the same question as Sarah Zaltash, 'Can't you see we are praying?'

On Maundy Thursday, the day which commemorates Christ washing of the feet of his disciples on the eve of his death, *Christians for Climate Action* organised a ceremonial foot washing for protestors around the site to 'serve, nurture and bless activists.' When police attempted to clear the Oxford Circus site, the protestors chained themselves to each other, forming a human barricade, an action which resulted in a number of arrests.

In explaining their commitment to non-violent civil disobedience, one member of the group, Holly-Anna Petersen noted one of their key aims was to 'challenge all Christians to question their cooperation with the system in which we live and to suggest to them that to be a Christian might mean non-compliance with ruling authorities. Nonviolent direct action isn't something that we go into impulsively, all guns blazing,' she added. 'It's something we need to be wise about, acknowledge the power structures at play and prayerfully consider how we can expose and invert these.' In 2018, Petersen worked with other *Christian Climate Action* supporters to drop a large banner off the balcony during a meeting of the *General Synod of the Church of England*. The banner read as follows:

> We are young Christians. For us and our children, climate change is the biggest threat we face. Please pray and act for all those afflicted by climate change now and in the future. As a church community, we cannot continue to invest in fossil fuel companies. So we ask you, on our behalf,

to divest now. May God defend the afflicted among the people and save the children of the needy'. (Is. 24:2–5)

Petersen speaks thoughtfully of her reaction the first time she heard the story of Jesus turning over the tables in the temple.

> I remember thinking Jesus isn't some push over, that is integrity and that is bravery. How many people here would be okay with doing the modern-day equivalent of that? Going into the building of the people who are the powerhouses of today, the oil companies or the banks that are funding them and turning over the tables, addressing the crowds and telling them about the corruption that they are causing?

In 2018, Greg Rolles, one of the founders of *Christian Climate Action Australia* was arrested for blocking the train line between the Adani coal mine and the Great Barrier Reef on the Queensland Coast. In convicting and fining him $75,000, the judge refused to accept Rolles' defence of acting legally and ethically to prevent the consequences of a clear and immediate climate emergency. Rolles explained the justification for his action in this way.

> I really don't want to be sued and lose everything I have, but I am more worried about global warming and the environmental injustices of this world. The greatest threat we face is global warming, and we have a responsibility as Christians especially in the Western world to non-violently disrupt and slow the process down for the sake of people in the majority world. Am I interested in answering to my own empire, or answering God's call to work for kingdom? [224]

These stories and examples all clearly reflect views at the more radical end of the spectrum of religious perspectives on the case for climate justice civil disobedience. In a world deluged with stories about unbridgeable chasms between the values and motivations of differing faith-based traditions, I find considerable food for thought in the equally strong evidence of shared concern about the fragility of creation and the urgency of decisive and courageous climate action. These concerns are also strikingly and increasingly apparent in the work of many Buddhist, Taoist and Confucian writers, teachers and activists.

8

This World Is but a Dew Drop World….and yet….: Buddhist, Taoist and Confucian Learning About Suffering, Impermanence and Compassion

When I mentioned to an environmental activist friend of mine that I was writing about Buddhist responses to the climate crisis, he smiled gently, wondering why I would bother. 'I can see' he said 'that Buddhist meditation practice might help you deal with personal pain and grief but I can't see it making much of a useful difference in helping most people deal with floods and fires in the real world. Nor in troubling the sleep of fossil-fuel corporation CEOs.' This interpretation of Buddhism as a philosophy of disengagement and a practice of passivity remains common even in some Buddhist circles.

In my experience, there are however other ways of understanding Buddhist teachings as a rich source of wisdom about ideas and practices with significant potential to inform and sustain thoughtful, compassionate and courageous climate action. This alternative, more engaged understanding of the implications of Buddhist principles for ecological politics and practice clearly informs the *2015 Buddhist Climate Change Statement to World Leaders* signed by the Dalai Lama along with 24 other senior Buddhist leaders [225].

> We believe it imperative that the global Buddhist community recognize both our dependence on one another as well as on the natural world. Together, humanity must act on the root causes of this environmental crisis, which is driven by our use of fossil fuels, unsustainable consumption patterns, lack of awareness, and lack of concern about the consequences of our actions.

I am also increasingly struck by the extent to which deeply committed, highly respected climate change campaigners and policy makers including former Californian Governor Gerry Brown, environmental activist and poet Gary Snyder and former United Nations Climate Change secretary Christiana Figueres are drawn to and sustained by Buddhist learning and insight.

Christiana Figueres speaks warmly, for example, of the role which the teachings of the contemporary Vietnamese Buddhist teacher Thich Nhat Hanh played in enabling her to maintain the emotional resilience, calm strength and sharp focus required for her to complete her leadership role in brokering the Paris Climate Agreement. 'I don't think' she reflects 'that I would have had the inner stamina, the depth of optimism, the depth of commitment, the depth of the inspiration if I had not been accompanied by the teachings of Thich Nhat Hanh….They were my guidance and the light in my life….and they also helped to maintain an inordinate amount of calm in moments of total crisis in the negotiations' [226].

Sitting Still and Sweeping the Garden: Insights and Pathways of Engaged Buddhism

Noting that some non-Buddhist readers find it difficult to relate to the unfamiliar and at times esoteric terminology of Buddhist writing, I have chosen to employ more everyday language, conscious also that readers with stronger backgrounds in Buddhist learning may find some of these reflections a little simplistic. An important starting point for any discussion of Buddhist, and indeed also of Taoist and Confucian teachings, is understanding that they are generally best read as philosophical rather than religious texts. Gautama Siddhartha, the man who became known as the Buddha, was a human being not a god. The teachings of the Buddha are therefore best understood as contestable guides to ways of experiencing and being in the world rather than as revealed truths to be taken on faith.

While the foundational story and key principles of Buddhism will be familiar to some readers, a brief recap may still be helpful. Gautama Siddhartha was born and grew up in a wealthy royal household on the border of India and Nepal around 550 BC. At the age of 29, Siddhartha became deeply troubled by his encounters with sickness, suffering, ageing and death. Following his decision to leave the royal palace, he spent six years seeking out the wisest teachers of spiritual wisdom and meditation practice before arriving at his own 'awakening' (the title Buddha simply meaning 'awakened' or 'enlightened'). The core ideas of the Buddha's 'awakened' teachings can be summarised in the following way.

Human beings experience many forms of suffering and distress in confronting the painful realities of ageing, illness, separation, loss and death. The original Sanskrit term *dukka* often imperfectly translated as 'suffering' is more accurately understood as 'that which is impossible to bear.' *Dukka* may also refer to the term for a poorly formed axle hole in a wheel, suggesting the source and difficulty of travelling smoothly and painlessly on the road of life.

From this perspective, the underlying cause of human suffering is the desperate, forlorn longing to achieve the impossible goal of preventing and controlling change and of acquiring and holding on forever to experiences and things which will inevitably pass away. The Buddhist pathway to overcoming human suffering and distress therefore leads us to focus on letting go of our unachievable desires for changelessness and certainty; of accepting and welcoming the inevitability of impermanence; and of understanding and respecting the ways in which all forms of life and indeed all matters are interconnected and interdependent.

'Nirvana,' the ultimate destination of the Buddhist path, is sometimes interpreted as 'heaven,' implying a world beyond our life on Earth. The original, literal meaning of 'Nirvana' 'mind like fire unbound' suggests an alternative interpretation of becoming fully aware of our true human nature, a flame burning brightly and shedding light, unbound to any material substance.

Key steps on the Buddhist pathway leading beyond suffering and despair include committing to and enacting the principles and practices of kindness and compassion; generosity and acceptance; honesty and truthfulness; meditation and contemplation, all underpinned by deep understanding of the impermanence and interdependence of our lives and of the worlds in which we live [227].

For some followers of Buddhist teaching, these principles are best enacted by relinquishing attachment to the world through a largely disengaged, meditative and perhaps monastic way of life. The decision to focus primarily on an individual meditative practice clearly provides one way of letting go of the pain of confronting and enduring the consequences of ecological and climatic collapse. Thich Nhat Hanh reflects in the following way on the journey which led him to explore and articulate an alternative pathway of socially engaged Buddhism.

> When I was in Vietnam, so many of our villages were being bombed. Along with my monastic brothers and sisters, I had to decide what to do. Should we continue to practice in our monasteries, or should we leave the meditation halls in order to help the people who were suffering under the bombs? After careful reflection, we decided to do both—to go out and help people and to do so in mindfulness. We called it engaged Buddhism. Mindfulness must be engaged. Once there is seeing, there must be acting. We must be aware of the real problems of the world. Then, with mindfulness, we will know what to do and what not to do to be of help. [228, p. 91]

Environmental activist and author of *Active Hope, How to Face The Mess We're in Without Going Crazy,* Joanna Macy suggests that social engagement is fully consistent with and flows logically from the Buddha's doctrine of 'dependent co-arising, the dynamic interdependence of all phenomena, [which aims] to liberate us from the prison cell of egocentricity, and from the greed, hatred, and delusion it engenders. Engaged Buddhism refers to the social application of these teachings, as they bring us into responsible relationship with the world around us' [229].

Buddhist author and scholar Jack Kornfield employs the metaphor of 'sweeping the garden' to describe our shared responsibility for caring for all the inhabitants of the world. 'There are only two things to do. Sit, and sweep the garden. This is like breathing in and breathing out. You quiet the mind and the heart so that you're connected to yourself and listen to what really matters. Then you get up from that stillness, and if people are hungry, you offer food. If there's injustice, you offer yourself for the healing of that injustice' [230].

An increasing number of socially engaged Buddhist practitioners are indeed 'rising from that stillness' to explore and follow Buddhist pathways which renew and embolden our commitment to climate action; deepen understanding of the underlying sources of ecological crisis; and strengthen our capacity to sustain action in the face of growing recognition of the harsh realities of the way our journey is likely to unfold.

People Who Love the Earth: Remembering Why We Care

Among the multitudes of claims and counter claims about the science and politics of emissions reduction and energy transition, Zen Buddhist scholar, poet and environmental activist Gary Snyder helps us recall the reason why so many people wake up every morning passionately committed to climate action. Asked to reflect on 'the things we need more of in Western societies,' Snyder's response is typically wide ranging and provocative.

> More women in politics; religious views which do not exclude nature and do not fear science; political leaders who have worked in schools, factories or on farms and write poems; intellectuals who studied history and ecology and who like to dance and cook; poets who do not care about literary criticism. But what we need most is people who love the Earth. [231]

Snyder's reflections on the need to constantly renew and celebrate our love for the Earth are underpinned by a deeply grounded ethical commitment to respect and care for the landscapes and ecologies of the remarkable planet on which we have had the good fortune to be born.

They also help us understand, as Thich Nhat Hanh notes us in his 2015 *Climate Change Statement to the United Nations*, that 'the Earth is not just our environment. The Earth is not something outside of us. Breathing with mindfulness and contemplating your body, you realise that you are the Earth.'

> We can all experience a feeling of deep admiration and love when we see the great harmony, elegance and beauty of the Earth. A simple branch of cherry blossom, the shell of a snail or the wing of a bat all bear witness to the Earth's masterful creativity. When you realize the Earth is so much more than simply your environment, you'll be moved to protect her in the same way as you would yourself. [227]

Much writing on the implications of Buddhist teaching for ecological activism begins by speaking of Siddhartha's original choice to meditate under a great forest tree in the company of animals and birds. These teachings often go on to speak about the legend of Siddhartha's response to the final assault on him by the army of the demon king, Mara. The story goes that Mara's soldiers began their attack shouting, 'who will bear witness for you? Who will stand with you?' Siddhartha placed his right hand gently and firmly on the Earth and the Earth itself roared back, 'I bear you witness!' At this moment, Mara and his army disappeared and Siddhartha awakened to the wisdom of the Buddhist path.

Buddhist literature includes a vast body of poetry honouring the mountains and rivers; oceans and forests of the Earth. Here, for example, is the way this gratitude is expressed in *The Therigatha*, a collection of poems written by Buddhist nuns between the sixth and third century BCE and generally regarded as the earliest known published collection of writing by women in India [232]. In reading these lines, it may be helpful to note that Indra, god of thunder, storms and rain, is one of the most important figures in both Buddhist and Hindu traditions. Many myths and stories imagine the clouds as Indra's cattle, with rain falling as the result of Indra's work in milking his celestial herds.

> Those rocky heights with hue of dark blue clouds
> Where lies embossed many a shining lake
> Of crystal-clear, cool waters, and whose slopes

The herds of Indra cover and bedeck.
Those are the hills wherein my soul delights.

I am also struck in rereading these lines from the *Cold Mountain* poems by the ninth-century Chinese Zen poet, Han Shan of the delight and comfort which human beings have shared for thousands of years in paying full and careful attention to the natural world [233].

As for me, I delight in the everyday Way.
Among mist-wrapped vines and rocky caves
Here in the wilderness I am completely free
With my friends, the white clouds, idling forever
There are roads, but they do not reach the world
Since I am mindless, who can rouse my thoughts?
On a bed of stone I sit, alone in the night
While the round moon climbs up Cold Mountain.

Overcoming Ignorance, Violence and Greed

In speaking about the forces which intensify suffering, Buddhist teachings foreground three particularly toxic psychological drivers: delusion and ignorance; aggression and violence; selfishness and greed. Many contemporary Buddhist writers point to the destructive roles which all three of these 'poisons' play in creating and reinforcing the underlying dynamics of climate change and ecological crisis.

The delusion that each human being is entirely separate and autonomous, disconnected and isolated from all other human beings and all other species is frequently employed as a convenient justification for the dominant 'greed is good' assumptions and values of neoliberal capitalism. The destructive consequences of this dangerous delusion are strikingly apparent in the abiding influence of the writings of Ayn Rand (1905–1982), recently described by Bill McKibben as arguably the most important political philosopher of our time. 'Indeed,' McKibben continues, 'given the leverage of the present moment, leverage that is threatening to end the human game, you could argue that she's the most important philosopher of all time.'

Rand's novels *The Fountainhead* and *Atlas Shrugged* are famously the preferred sources of guidance and inspiration for many of the most powerful masters of the neoliberal universe including Steve Jobs, Allen Greenspan, Uber's founding CEO, Travis Kalanick, Rex Tillerson and Mike Pompeo [234]. Donald Trump describes himself as a huge Ayn Rand fan, identifying closely with Rand's hyper-individualistic hero, Howard Roark. *The Fountainhead,* in Trump's view, 'relates to business, beauty, life and inner emotions. That book relates to…..everything' [235].

Rand herself summarised the core values of her philosophy as the belief that 'man exists for his own sake, that the pursuit of his own happiness is his highest moral purpose, that he must not sacrifice himself to others, nor sacrifice others to himself.' 'By my life and my love of it' vows one of her most famous characters, 'I will never live for the sake of another man, nor ask another man to live for mine' [236, p. 1069].

It is easy to see how a political and economic system dominated by individuals with such extreme individualistic values could continue to drive forward into a future of accelerating inequality, unconstrained consumerism and ecological risk with such alarming speed. It's also not surprising that the current head of the *Ayn Rand Institute,* Yaron Brook remains such an influential advocate for the view that climate science and climate activism are part of a sinister narrative skilfully employed by socialists and environmentalists to expand the role of the state and undermine individual freedom.

More broadly and as the eminent Tibetan Buddhist teacher Ringu Tulku Rinpoche notes, the belief in the dualistic view that there is a separate 'I' that is not part of anything else also creates fertile ground for aggression and violence against other human beings and other species.

> If there is a separate 'I' then there must be separate 'others.' Up to here is 'me.' The rest is 'they.' As soon as this split is made, it creates two opposite ways of reaction: 'This is nice, I want it!' and 'This is not nice, I do not want it!' On the one hand there are those things that seem to threaten or undermine us. Maybe they will harm us or take away our identity. They are a danger to our security….Then on the other hand there are those

things that are so nice. We think, 'I want them. I want them so much.' Through this way of thinking attachment arises. [237, p. 29]

American author David Loy, who has written widely on engaged Buddhist responses to the climate crisis, usefully illustrates the way in which the three 'toxins' of delusion, violence and greed have become institutionalised [238]. The illusion that there are no longer any viable alternatives to competitive individualism and unconstrained consumerism has become deeply embedded as taken for granted, common sense in many cultures. State-sanctioned violence and brutality have become default policy options for many governments in dealing with refugees and asylum seekers. And the formidable resources and techniques of the advertising and marketing industry continue to entrench acceptance of the necessity of endlessly expanding consumerism and economic growth.

Loy points us to the work of Austrian satirist Karl Kraus in speaking about ways in which the 'one per cent' responsible for maintaining dominant neoliberal power structures in fact suffer from the same self-destructive delusions that increasingly infect us all. 'How do wars begin? Politicians tell lies to journalists, then believe what they read in the newspapers. The same applies to shared fantasies such as the necessity of consumerism and perpetual economic growth, and collective repressions such as denial of impending eco-catastrophe' [239].

Interdependence and Impermanence; Compassion and Generosity

From a Buddhist perspective, the antidotes to delusion, violence and greed—deepening awareness of the interdependence and impermanence of the material world and deepening respect for practices of compassion and generosity—provide many valuable insights into ways of focusing and sustaining action in dark times. Viewing our own lives and the lives of all other human beings as interconnected and interdependent rather than as wholly separate and autonomous can open our eyes to ways of living and working which prioritise reciprocity, co-operation and well-being. 'Everything,' Zen Buddhist teacher Robert Aitken suggests, 'is contingent upon everything else.'

> Plants transpire, the moisture evaporates and returns as rain. The Earth is dampened, allowing rootlets to absorb nutrients in the soil. The nutrients themselves are released by worms that eat the Earth, and by the casts of countless other beings as they give themselves in death. People, animals, and other plants flourish, and give themselves in turn. [240]

As UCLA Berkeley economics Prof Claire Brown notes in her introduction to *Buddhist Economics, An Enlightened Approach to the Dismal Science*, a world view framed through the lens of interdependence has significant implications for the ways in which we theorise and implement economic policies. 'Interdependence in Buddhist economics' Brown argues 'is expressed in three ways. The first involves using resources to enhance the quality of life for ourselves and for others. The second integrates caring for Nature and our environment into all activities. And the third involves reducing suffering and practicing compassion, both locally and globally' [241].

Small is Beautiful: A Study of Economics As If People Mattered, first published by E.F. Schumacher in 1973 played a key role in introducing the principles of Buddhist economics to Western audiences. Schumacher argues that a thoughtful and well-informed understanding of ecological interdependence is an essential foundation for a just and sustainable transition to a zero-carbon economy.

> Just as a modern European economist would not consider it a great achievement if all European art treasures were sold to America at attractive prices, so the Buddhist economist would insist that a population basing its economic life on non-renewable fuels is living parasitically on capital instead of income. As the world's resources of non-renewable fuels – coal, oil and natural gas are exceedingly unevenly distributed over the globe and undoubtedly limited in quantity it is clear that their exploitation at an ever-increasing rate is an act of violence against nature which almost inevitably lead to violence between men. [242, p. 45]

This broader and more holistic view of the purpose of economics has led to growing interest in ways of assessing national 'progress' based on integrated measures of social, economic and ecological wellbeing rather than the one-dimensional metric of Gross Domestic Product. The Gross National Happiness (GNH) wellbeing framework developed by the government of Bhutan and explicitly informed by Buddhist values takes account, for example, of the extent to which people have time for meditation and to care for their children and old people alongside improvements in income, health and education.

In 2011, I had the honour of visiting Bhutan to attend a remarkable conference on *Happiness, Wellbeing and Human Development* convened by the Prime Minister of Bhutan, the Hon. Jigme Tinley. In addressing the conference, Prime Minister Thinley spoke passionately about the implications for climate action of a development paradigm informed by far deeper understanding of the interconnections between sustainability, equity, human values, ecological resilience and good governance. 'At COP 15 in Copenhagen' Prime Minister Thinley explained 'Bhutan pledged to remain a net carbon sink in perpetuity. This, and our belief in the web of interdependence among all life forms, has led to our biodiversity and watershed protection measures, our wildlife corridors, and the many other ways in which the natural environment is placed at the very core of all our development policies' [243].

English economist and author of *Doughnut Economics* Kate Raworth also draws on Buddhist insights about social, ecological and economic interdependence in developing her argument for creating and sustaining a 'safe and just space for humanity' (and other species) through the creation of a regenerative and distributive economy [244]. 'Humanity's 21st century challenge' Raworth argues 'is to meet the needs of all within the means of the planet....while ensuring that collectively we do not overshoot our pressure on Earth's life-supporting systems, on which we fundamentally depend....The Doughnut of social and planetary boundaries is a playfully serious approach to framing that challenge, and it acts as a compass for human progress this century' [245].

The outer circle—or 'ecological ceiling' of Raworth's Doughnut diagram—consists of nine 'Planetary Boundaries' essential for human survival and flourishing: climate change, ocean acidification, chemical pollution, nitrogen and phosphorous, freshwater withdrawals, land conversion, biodiversity loss, air pollution and ozone depletion [246]. The inner circle of the Doughnut is made up of twelve foundational social dimensions. These dimensions, closely aligned with and informed by the *UN Sustainable Development Goals,* include water, food, health, education, income and work, peace and justice, political voice, gender equity, housing, networks and energy. The safe and just space for humanity to survive, the space for a regenerative and distributive economy lies Raworth suggest in the space between the planetary boundary ceiling and the social wellbeing foundation.

In April 2020, the City of Amsterdam formally adopted the *Doughnut Economics* framework as the basis for developing and implementing just and ecologically sustainable pandemic recovery policies [247]. In explaining the relevance of the Doughnut framework to pandemic recovery and the climate crisis, Raworth noted that 'the world is experiencing a series of shocks and surprise impacts which are enabling us to shift away from the idea of growth to thriving. Thriving means our wellbeing lies in balance. We know it so well in the level of our body. This is the moment we are going to connect bodily health to planetary health' [247]. UN Secretary General, Antonio Guterres clearly had similar principles in mind when speaking to the *G20 Leaders Group* about pandemic recovery priorities in March 2020.

> Finally, when we get past this crisis – which we will – we will face a choice. We can go back to the world as it was before or deal decisively with those issues that make us all unnecessarily vulnerable to crises....The recovery from the COVID-19 crisis must lead to a different economy. Everything we do during and after this crisis must be with a strong focus on building more equal, inclusive and sustainable economies and societies that are more resilient in the face of pandemics, climate change, and the many other global challenges we face. [248]

Awareness of impermanence, the realisation that all our lives are just 'a flash of lightening in a summer cloud, a flickering lamp, a phantom and a dream,' is sometimes understood as an argument for disengaging from the pain and suffering of the illusory, material world. Thich Nhat Hanh notes however that the deeper challenge arising from the understanding that 'all things pass' is to learn the art of holding together two apparently contradictory ways of being in the world: to relinquish our attachment to the ultimately futile goal of controlling the uncontrollable while at the same time acting with compassion, kindness and generosity to reduce suffering in the times and places in which we currently exist.

> All civilisations are impermanent and must come to an end one day. But if we continue on our current course, there's no doubt that our civilisation will be destroyed sooner than we think. The Earth may need millions of years to heal, to retrieve her balance and restore her beauty. She will be able to recover, but we humans and many other species will disappear, until the Earth can generate conditions to bring us forth again in new forms.
>
> Once we can accept the impermanence of our civilization with peace, we will be liberated from our fear. Only then will we have the strength, awakening and love we need to bring us together. Cherishing our precious Earth–falling in love with the Earth–is not an obligation. It is a matter of personal and collective happiness and survival. [227]

Roshi Joan Halifax, Abbot and head teacher at the *Upaya Zen Centre* in Santa Fe, New Mexico, helps us understand that realising that the only certainty is change can also assist us in sustaining courageous action in full knowledge that all our efforts may not, in the end, create the outcome we are seeking. In reflecting on her many years of working on the seemingly hopeless causes of refugee and climate activism and on her work as a caregiver for terminally ill patients and prisoners on death row, Halifax wonders aloud.

> Why work in such hopeless situations? Why care about ending the direct and structural violence of war or injustice, as violence is a constant in our world? Why have hope for people who are dying, when death is inevitable; why work with those who are on death row...or serve refugees fleeing from genocide, and no country seems to want these men, women, and children? [249]

Halifax's answer foregrounds the importance and the difficulty of 'showing up'; of continuing to act with compassion and generosity, knowing full well that all the human beings and other species we are caring for will one day disappear.

> Sitting with a dying person or a dying planet, we show up, we do the best we can, we rely on altruism, empathy, integrity, respect, engagement, and most importantly compassion and wise hope…I learned that I had to do my best by moving away from the story that working for peace, justice, or an equitable and compassionate society.…would turn out well, was too big a job, or was hopeless. At the same time, I could not be attached to any outcome, as I intuitively knew that futility might paralyze me. I had to 'just show up' and do what I felt was morally aligned with my values, my principles, my commitments, regardless of what might happen. [250]

The relevance and significance of these various Buddhist teachings and insights for the ecological and political crises now unfolding around us are thoughtfully expressed in the following reflections by Gary Snyder.

> It may well be that it's already far too late to have any effect on the progress of climate change and its effect on ecosystems and human populations. Although alternative energy resources work in specific cases and places, they cannot stand in for the energy demands that will keep the global economy from making more nuclear plants, drilling for more oil and gas and mining for more coal.…..Truth is, we all live right now under the shadow of a great and intractable empire, the Global Economy—capitalism with no roots or grounding anywhere, dedicated to making profits until it all collapses.
> Yet, still, every day, I feel gratitude to this world, that is, Issa's haiku goes: 'This dewdrop world is but a dewdrop world… and yet….' [21, p. 15]

People Follow Earth, Earth Follows Heaven, Heaven Follows the Way: Taoist Wisdom in a Time of Climate Crisis

Lao Tzu, so some stories go, was a Chinese philosopher in the royal court during a period of bitter war and conflict in the sixth century BCE. Deeply disillusioned by his experience of widespread corruption, he decided to go into exile and live as a hermit. While passing through the gateway out of the kingdom, the guard, recognising him as a great philosopher, asked him to write a book before he left so that his wisdom would not be lost and could be shared with future generations. The contents of this book became the *Tao Te Ching*, 'the book of the Way,' the second most widely read book after the Bible [251]. Some writers also say that Lao Tzu then travelled south to India and became the teacher of the young Siddartha Gautama.

Remembering that the opening line of the *Tao Te Ching* is often translated as 'the Tao that can be described is not the real Tao,' we can perhaps speak cautiously and imperfectly of the Tao as a pathway and a guide through which we discover and follow the patterns, rhythms and flows of the universe. For Allan Watts, the English philosopher responsible for bringing Taoist and Buddhist ideas to western audiences in the 1960s, the Tao can also be understood as a way of learning to cooperate 'with the course or trend of the natural world, whose principles we discover in the flow patterns of water, gas, and fire, which are subsequently memorialized or sculptured in those of stone and wood, and, later, in many forms of human art...' [252, p. xiv]. More lyrically and elusively, the second century BCE Taoist text, *The Huainanzi* refers to the Tao in the following way.

> Mountains are high because of it.
> Abysses are deep because of it.
> Beasts can run because of it.
> Birds can fly because of it.
> The sun and moon are bright because of it.
> The stars and timekeepers move because of it. [253, p. 17]

While some readers may find these ideas a little esoteric, Taoist scholars and environmental activists Chen Xia and Martin Schonfeld suggest several more tangible implications of Taoist ideas for policy and practice in times of climate crisis [254]. 'The Tao brings forth life and complexity, and humans should do the same.'

> Climate change is a visceral reminder that we are in the world; that our being is part of a larger being, and that going against the flow of nature means to run counter our collective self-interest....Our first responsibility in a changing climate is to learn to become stewards of the biosphere, to shepherd life, and protect complexity. [254, p. 202]

The poetry of the *Tao Te Ching* is deeply imbued with respect for the interdependence of human beings and all other species; of rivers and mountains; Earth and sky. 'People follow Earth, Earth follows heaven, heaven follows the Way....Love the world as your own self; then you can truly care for all things.' Taoist principles also suggest that the number and diversity of species might be a better way of judging the affluence of a society than GDP. 'If all things in the universe grow well, then a society is a community of affluence. If not, this kingdom is on the decline' [255].

From the Taoist point of view, human beings have a clear and abiding responsibility to help maintain the delicate balance, complex flows and intricate relationships of our world.

> In harmony with the Tao, the sky is clear and pure,
> The Earth is serene and whole,
> The spirit is renewed with power
> Streams are replenished
> The myriad creatures of the world flourish, living joyfully,
> Leaders are at peace and their countries are governed with justice
>
> When humanity interferes with the Tao,
> The sky turns filthy, the Earth is depleted
> The spirit becomes exhausted
> Streams run dry, the equilibrium crumbles
> Creatures become extinct [251]

The *Tapingjing*, a 2000-year-old collection of Taoist writing focusing on the sources and dangers of ecological and civilisational crisis, provides a strikingly prescient warning of the consequences of failing to maintain the balance of the world. 'The heavenly signs and earthly patterns are in disorder, and thus heaven and Earth become ill. This makes the sun, moon, and stars, wind and rain, the four seasons fight. It will be an inauspicious year for crop harvest' [256, p. 356].

Taoism, Xia and Schonfeld suggest, also provides an alternative model of being-in-the-world. 'The Tao acts by non-acting; that is, by acting in harmony with the natural flow. The second responsibility [of humans] is, accordingly to refrain from further disruptions of the flow, and to learn to become mitigators of climate change, to soften the impact and to calm down the waves' (p. 200).

> Ruling the country is like cooking a small fish....
> One must know when to stop.
> Knowing when to stop averts trouble.
> Tao in the world is like a river flowing home to the sea. [251]

The Taoist principle of non-action, sometimes interpreted as 'doing nothing,' can also be seen as an argument for 'effortless or non-calculating action'; as the opposite of acting in a compulsive, reckless and wilful way'; of 'acting in a way in which nothing that matters is left undone.' There are also potentially significant insights here supporting and informing the politics and strategies of non-violent civil disobedience [254].

> Under heaven nothing is more soft and yielding than water
> Yet for attacking the solid and strong nothing is better. It has no equal
> The weak can overcome the strong.
> The supple can overcome the stiff. [251]

Others have noted the implications of Taoism for building an economics of voluntary simplicity, of treading lightly on the Earth.

> Better stop short than fill to the brim
> Over sharpen the blade and the edge will soon blunt.

> Amass and store of gold and jade and no one can protect it.
> Claim wealth and titles and disaster will follow.
> He who is attached to things will suffer much
> He who saves will suffer heavy loss
> A contented man is never disappointed. [242]

And others again have noted the synergies with Buddhist teachings about the importance of kindness, sincerity, compassion and generosity.

> In dwelling live close to the Earth
> In matters of the heart, seek depth
> In relationships be kind and generous
> In speaking be truthful and sincere
> In leading be just and fair
> In work strive to be competent
> In acting remember that proper timing is everything. [251]

All Under the Heavens Belongs to the Public: Ecological Intelligence in Confucian Thought

The sixth century BC 'Warring States' period of Chinese history, during which Confucius lived and wrote aligns closely with the life of Lao Tzu and Siddhartha Gautama. The Confucian tradition has much in common with Taoism, viewing Heaven, Human beings and Earth as a trinity of worlds or spheres interconnected through the flows and patterns of the Tao. While Confucian teachings often place strong emphasis on improving the material wellbeing and livelihoods of human beings, this does not mean that the natural world should simply be regarded as a resource to be exploited solely for human benefit.

Many Confucian authors employ metaphors of the family and the body to illuminate the depth and complexity of the relationship between human and non-human worlds. Eleventh-century neo-Confucian philosopher Zhang Zai expresses this relationship in the following way. 'Heaven is my father and Earth is my mother and even such a small creature as I find an intimate place in their midst. Therefore that which fills the universe I regard as my body and that which directs the universe I consider as my nature. All people are my brothers and sisters, and all things are my companions' [242, p. 107].

Fifteenth-century Chinese philosopher Wang Yangming speaks even more directly about the ways in which Confucian ideas evolved in very different directions to the more dualistic perspectives articulated in Western philosophical and theological circles. 'The great man regards Heaven and Earth and the myriad things as one body. He regards the world as one family and the country as one person. As to those who make a cleavage between objects and distinguish between self and others, they are small men' [248].

An alternative metaphor frequently employed by Confucian authors to describe the relationship between Humans, Earth and Heaven is the image of the ship (of human civilisation) sailing on the waves of a vast and stormy ocean (the natural world). This image is sometimes contrasted with the Western viewpoint of human civilisation as a great land mass (created by God) surrounding and drawing on the resources of an infinitely exploitable ecological lake.

Confucian thought initially emphasised the principle that the 'Mandate of Heaven' was confined to the Emperor, a concept similar in its implications to the Western principle of the Divine Right of Kings. The crucial shift in neo-Confucian thought was to replace the Mandate of Heaven with the more egalitarian and democratic concept of 'Tianxiaweigong'......'All under heaven belongs to the public' [249]. If all under heaven does indeed belong to the public, then the legitimacy of rulers and governments depends on their capacity to be responsive to the concerns and priorities of the people.

Sang-Jin Han, Professor of Sociology at Seoul National University who has played a key role in building bridges between Taoist and Confucian ideas and the work of Western critical social theorists including Jurgen Habermas and Ulrich Beck, argues that the way we interpret the concept of 'all under heaven belongs to the public' has important implications for the ways in which we address climate change and related ecological challenges. Han contends that there are three main ways in which the interests and priorities of 'the public' can be understood [250].

The first way of framing and pursuing the 'the public interest' emphasises the goal of improving human security and material wellbeing, ideally in ways which are as just and inclusive and fair as possible. This aim of overcoming poverty and strengthening overall living standards arguably remains the dominant underlying motivation of the socialist modernisation project of contemporary China.

The second approach to promoting the public interest places greater stress on the democratic principle of protecting opportunities for the voices and views of individuals to be expressed, discussed and heard. Professor Han notes the close alignment between this perspective and Jurgen Habermas' arguments for strengthening the institutional foundations and processes of communicative action and deliberative democracy.

The third way of understanding the public interest is through an ecological lens, focusing more closely on the escalating risk of climatic and ecological collapse and the futility of prioritising human wellbeing on a planet incapable of sustaining human life.

The difficulty of reconciling the tensions between these three materialistic, democratic and ecological ways of understanding and promoting the interests of human beings leads us back to our image of the metaphorical ship of human civilisation navigating an ocean swept by increasingly severe climatic storms. Keeping the civilisational ship afloat will clearly require ensuring the crew are sufficiently contented and well fed not to mutiny. Communication and decision-making among the diverse members of the crew must continue to be respectful and incisive. And perhaps most importantly of all the crew must also be deeply attentive to the ever-increasing power of the wind and the height of the waves.

There are grave dangers, Han argues in overemphasising any one of these approaches. An overemphasis on the sensitivities and nuances of deliberative democracy might, for example, prevent the crew from taking action at the speed required to prevent being overwhelmed wrecked. On the other hand, there are also clear risks in creating an ecological totalitarian regime under the banner of the ecological state or perhaps even of a 'climate emergency.'

One way of navigating these complex choices, Han suggests, may be to consider the meaning of 'jen' the highest virtue in Confucian thought. In Confucian teaching, 'jen' is the quality which gives human beings their humanity, encompassing and integrating Buddhist, Taoist and Confucian principles of interdependence, kindness, compassion and generosity. 'Jen,' Han argues, is 'the desire to preserve and fulfil one's own life and also to preserve and fulfil the lives of others.'

> Inhumanity (the absence of jen) is the desire to preserve and fulfil one's own life to the point of destroying the lives of others. The man or woman of jen regards Heaven and Earth and all things as one body. To him there is nothing that is not himself. Insofar as he recognises all things as himself, can there be any limit to his humanity? To be charitable and to assist all things is the foundation of a sage. [261]

This central idea, that it makes no sense to think of ourselves as wholly independent, isolated individuals; that no one, no man or woman is an island has seemed self-evident to me for as long as I can remember. In rereading John Donne's reflections on interdependence, solidarity and reciprocity, I am also struck by the ongoing relevance of these ideas to the creation and communication of the ecologically informed ways of being we will require to survive and flourish in a world of rising oceans and of constantly escalating fires, floods and storms.

> No man is an island, entire of itself;
> every man is a piece of the continent, a part of the main.
> If a clod be washed away by the sea,
> Europe is the less, as well as if a promontory were;
> as well as if a manor of thy friend's or of thine now were.
> Any man's death diminishes me because I am involved in mankind;
> and therefore never send to know for whom the bell tolls; It tolls for thee.
> [262, p. 109]

9
Living Ecologically: Understanding and Respecting Complexity and Fragility

Creating and sustaining lives of meaning and purpose in an increasingly harsh climate will require human beings to learn to think and act in very different ways to the narrowly individualistic and short-sighted paradigms which have reflected and protected the interests of the wealthiest and most powerful individuals, classes and corporations over the last few hundred years. We cannot, as Albert Einstein famously pointed out, 'solve problems by using the same kind of thinking we used when we created them.' This chapter focuses therefore on the work of eight writers whose work I have found particularly helpful in deepening our understanding of the world as an intricate, interdependent pattern of fragile and constantly changing ecologies: Rachel Carson, Gregory Bateson, Donella Meadows, Arne Naess, Tim Morton, Bruno Latour and Val Plumwood.

So Delicately Interwoven Are the Relationships...

While, as previously noted, Rachel Carson's life and work were inspired and sustained by her delight in the wonders and mysteries of nature, her research and writing were also informed by rigorous application of the scientific and political implications of ecosystem ecology. 'It is useless' Carson reflected 'to attempt to preserve a living species unless the kind of land or water it requires is also preserved. So delicately interwoven are the relationships that when we disturb one thread of the community fabric we alter it all — perhaps almost imperceptibly, perhaps so drastically that destruction follows' [68, p. 256].

In 2012, in an article celebrating the 50th anniversary of *Silent Spring*, Canadian author Margaret Atwood revisits Carson's cautionary advice about the dangers of pursuing simplistic, short-term strategies for managing complex ecological systems. These risks, Atwood warns, are being intensified by our failure to recognise the ecological interdependence and fragility of our own physical existence. 'The inside of your body is connected to the world around you, and your body too has its ecology, and what goes into it–whether eaten or breathed or drunk or absorbed through your skin – has a profound impact on you' [263].

For Carson, the great tragedy of our human condition lies in the destructive consequences for so many species of insects, birds and animals—and ultimately ourselves—of our abiding faith in the omniscience and omnipotence of human technological wizardry. 'The "control of nature" is a phrase conceived in arrogance, born of the Neanderthal age of biology and philosophy, when it was supposed that nature exists for the convenience of man....It is our alarming misfortune that so primitive a science has armed itself with the most modern and terrible weapons, and that in turning them against the insects it has also turned them against the Earth' [67, p. 257].

Careful observation of meteorological and ecological data also deepened Carson's understanding of the underlying causes of these disturbing trends. As early as 1951, she noted that 'the evidence that the top of the world is growing warmer is to be found on every hand. The recession of the northern glaciers is going on at such a rate that many smaller ones have already disappeared. If the present rate of melting continues others will soon follow them' [264].

The conceptual framework underpinning *Silent Spring* was built on a series of foundational debates in the first half of the twentieth century about competing models of ecosystem ecology. The core assumptions informing these debates continue to inform contemporary responses to the social and ecological risks of the climate emergency. In 1935, the leading protagonist on one side of this conflict, the British botanist and zoologist, Authur Tansley coined the term 'ecosystem science' in referring to the complex and constantly evolving interdependence of ecological communities [265, p. 268]. In doing so, Tansley was aiming to critique and discredit the dominant ecological paradigm of the time, known as 'holism' popularised and championed by the South African botanist (later Prime Minister of South Africa and leading architect of apartheid), Jan Christian Smuts.

Smuts' paradigm of 'holism' emphasised the ways in which self-regulating ecosystems are maintained and stabilised by ensuring that each organism, each species and, in his view each race play their appropriate, pre-ordained role. Smut's proposition that 'healthy' well-functioning ecologies and societies require some organisms to play a more 'exalted' role than others continues to provide a convenient justification for political programs based on oppressive and hierarchical power relations. This doctrine of 'holism' also aligns closely with world views locating the human species at the apex of evolutionary and ecological pyramids, with some members of the species (white males for example) playing a particularly 'exalted' role. Tansley, well aware of the potentially dangerous political implications of Smut's work, articulated a very different view of ecological relationships and dynamics.

> Though the organisms may claim our primary interest we cannot separate them from their special environment, with which they form one physical system. These ecosystems, as we may call them, are of the most various kinds and sizes. They form one category of the multitudinous physical systems of the universe, which range from the universe as a whole down to the atom. [266, p. 521]

Tansley's carefully nuanced and ethically informed approach to ecosystem science inspired and informed the work of two other eminent ecologists, Charles Elton and Robert Rudd whose work directly influenced the writing of *Silent Spring*. In 1958, Charles Elton, founding Director of the *Bureau of Animal Ecology* at Oxford University, drew on his groundbreaking research on *The Ecology of Invasions by Animals and Plants* to deliver a stinging critique of the dangers of the unregulated use of synthetic pesticides. In Elton's view, 'this astonishing rain of death upon so much of the world's surface' threatened to overwhelm and destroy 'the very delicately organized interlocking system of populations.' Perhaps, Elton noted, we might one day look upon the consequences of our unthinking use of 'chemical warfare' to control and destroy insects in the same way in which we now understand 'the excesses of colonial exploitation' [267, p. 142].

These words seem particularly prescient in 2020 as protests against the neo-colonial racism of police brutality against Afro-Americans erupt across the United States. *Black Lives Matter* activist Leah Thomas also provides the following sharp and timely analysis of the importance of understanding the linkages between racial and environmental justice. 'The longer racism is not addressed the harder it will be to save the planet…..every environmentalist needs to hold themselves accountable and do the inner anti-racism work to achieve both climate and social justice…..'

> Intersectional environmentalism is an inclusive version of environmentalism that advocates for both the protection of people and the planet. It identifies the ways in which injustices happening to marginalized communities and the Earth are interconnected. It brings injustices done to the most vulnerable communities, and the Earth, to the forefront and does not minimize or silence social inequality. [268]

Carson also drew extensively on American ecologist Robert Rudd's critique of the tendency for economic systems focused on short-term profits to prioritise ecologically simplified crop systems requiring constantly escalating applications of insecticides. 'The modern world worships the gods of speed and quantity, and of the quick and easy profit, and out of this idolatry monstrous evils have arisen' [269, p. 31]

'We stand now,' Carson concluded, 'where two roads diverge. But unlike the roads in Frost's familiar poem, they are not equally fair. The road we have long been traveling is deceptively easy, a smooth superhighway on which we progress with great speed, but at its end lies disaster. The other fork of the road – the one less travelled by – offers our last, our only chance to reach a destination that assures the preservation of the Earth' [67, p. 240].

Pathways to an Ecology of Mind

My first reading, in the early 1970s, of *Silent Spring* left me with a host of questions about the guides and signals leading us onto this potentially disastrous superhighway. I was therefore delighted and intrigued to discover Gregory Bateson's wide-ranging studies in anthropology, zoology, cybernetics, linguistics and psychology collected together in *Steps to an Ecology of Mind*, first published in 1972. 'The major problems in the world' Bateson concluded in reflecting on the consequences of our failure to adequately respect ecological diversity and fragility 'are the result of the difference between how nature works and the way people think' [270].

> If you put God outside and set him vis-à-vis his creation and if you have the idea that you are created in his image, you will logically and naturally see yourself as outside and against the things around you. And as you arrogate all mind to yourself, you will see the world around you as mindless and therefore not entitled to moral or ethical consideration. The environment will seem to be yours to exploit. If this is your estimate of your relation to nature and you have an advanced technology, your likelihood of survival will be that of a snowball in hell. You will die either of the toxic by-products of your own hate, or, simply, of overpopulation and overgrazing. [271, p. 436]

For Bateson, the fatal flaw in dominant assumptions about the relationship between human beings and nature is the view that the Earth is best understood as no more than a vast pile of discrete, inert and unrelated objects. With characteristically dry humour, Bateson observed that he has himself sometimes been infected with this virus....'There are times when I catch myself believing there is something which is separate from something else' [272]. The consequences of this misalignment between the way some people (particularly the wealthiest and most powerful) think and the fragile interconnections of the world which many of us experience are closely aligned with the assumption that all the objects in the world are simply a treasure trove of resources waiting to be exploited and consumed.

Bateson also noted the consequences of the widely held belief that the human mind exists entirely separately from the world. This leads many people to mistake 'the map' (the interpretive frameworks we use to select the elements of the world important to us) for 'the territory,' the intricate ecological and cultural patterns which we can only ever know at best imperfectly. American environmental activist Chet Bowers draws on this insight in his comments on risks resulting from our consistent failure to recognise the selective processes through which we organise and prioritise our perceptions. 'If only the increase in profits is given attention when planting genetically modified seeds the differences which make a difference that signal environmental damage will go unnoticed. Concern with the loss of employment may lead to ignoring the carbon dioxide that the industry releases into the atmosphere, and that is contributing to changes in the ocean's chemistry' [273, p. 13]

Bateson's influential concept of 'the double bind,' originally formulated to help explain family dynamics and conflicts, also provides a useful metaphor for explaining our tendencies to avoid and deny the evidence and implications of climate science. The 'double bind' refers to a situation in which an individual receives two contradictory messages, with the second message operating at a higher, often less transparent level. A child might, for example, find it difficult and painful to follow through on her father's instruction to always tell the truth, if she learns that honestly sharing her political views with him leads rapidly to the withdrawal of parental love and approval. In a similar vein, many of us are tempted to close our eyes and minds to the consequences of choosing the short-term benefits of fossil-fuel-dependent lifestyles and economies over the actions required to achieve emergency speed emission reductions. In both cases, the immediate temptation to repress and conceal the truth has the potential to create longer-term pathologies.

Bateson's observations about the limited capacity of the conscious mind to fully grasp the intricate complexities of the world lead him to be extremely cautious of superficially appealing techno-fixes such as the unconstrained and unregulated use of pesticides, GMOs and geoengineering. His alternative paradigm prioritises a more humble, more nuanced 'ecological intelligence' informed by greater respect for the ways in which all aspects of the world, including the human mind, are interrelated and interdependent.

> We social scientists would do well to hold back our eagerness to control that world which we so imperfectly understand. The fact of our imperfect understanding should not be allowed to feed our anxiety and so increase the need to control. Rather our studies could be inspired by a more ancient, but today less honoured, motive: a curiosity about the world of which we are part. The rewards of such work are not power but beauty [271, p. 240]

Bateson's emphasis on the comforts and rewards of beauty over power is no accident. Aesthetic sensitivity cultivated and intensified through poetry and painting; music and dance; film and drama can, Bateson believed, open the doors of our perception to patterns and tapestries far richer than the data-driven outputs of an overly narrow focus on scientific rationality alone. In ways very similar to Rachel Carson, Bateson drew great delight and comfort from his time in the mountains and forests; deserts and oceans of the Earth. 'In the end,' he reflected 'the pathology of wrong thinking in which we all live....can only be corrected by an enormous discovery of those relationships which make up the beauty of nature' [270].

There Is Such a Thing as Enough

My initial responses to reading *Limits to Growth*, the ground breaking 1972 MIT research report on sustainability risks and challenges coauthored by Donella Meadows, also remain strikingly clear [59]. Here, I thought was the carefully considered evidence which would surely convince even the most sceptical critics of the urgency of recognising that infinite growth on a finite planet was a physical impossibility. The report's conclusions about the desirability and feasibility of more sustainable and integrated approaches to social and ecological wellbeing also seemed eminently sensible 'The state of global equilibrium could be designed so that the basic material needs of each person on Earth are satisfied and each person has an equal opportunity to realize his individual human potential' [59].

The sustained ferocity of the arguments and forces mobilised to derail broad acceptance and implementation of the key ideas of *Limits to Growth* contributed to Meadows' lifelong commitment to identify tools and strategies for accelerating well-informed and well-considered system-wide transformations. Many of Meadows' insights about strategies for imagining and inspiring transformational change remain highly relevant as we confront the harsh realities of a world in which much of the wise counsel contained in *Limits to Growth* has largely been ignored.

> Speak the truth. Speak it loud and often, calmly but insistently, and speak it, as the Quakers say, to power. Material accumulation is not the purpose of human existence. All growth is not good. The environment is a necessity, not a luxury. There is such a thing as enough. [274]

In *Leverage Points: Places to Intervene in a System*, Meadows identified a series of 'leverage points' for achieving system-wide change, organised in ascending order of impact and effectiveness (see App. A) [275]. Her view of these leverage points is informed by her understanding that the sustainability of any living system—river or forest; human body or city—depends on maintaining a healthy relationship between stocks (the state of the system at a particular moment) and flows (the factors influencing the system over time).

The quantity and quality of water in a river or lake can, for example, be altered by setting targets and standards for water inflows, usage and purity. We can also build larger dams or invest in better channels, pipes and pumps. All these interventions will however only have limited, short-term impacts and will finally come up against larger system constraints unless they are set at levels sufficiently high to trigger long-term behavioural and system-wide changes.

Policies for achieving broader system-wide transformations include negative feedback loop interventions such as price signals designed to reduce water consumption or pollution taxes to reduce the discharge of toxic chemicals. Positive feedback loops for a water source can be managed by controlling the amount of chemical fertilisers and nutrients entering a lake. While the expansion of nutrients will initially stimulate the growth of plants and fish numbers, if left unchecked the ultimate impact will be the exhaustion of oxygen and the extinction of life in the lake. Slowing down positive feedback loops like nutrient flows—or population or economic or emissions growth—can often have a more powerful impact on ecological sustainability than purely negative feedback interventions such as pollution or consumption taxes.

One of Meadow's sharpest insights is her recognition that levers of change become more powerful the closer we come to focusing on access to information flows and to the setting and enforcement of system rules. 'Power over the rules is real power....If you want to understand the deepest malfunctions of systems, pay attention to the rules, and to who has power over them' [275, p. 158]. The capacity to change the systems through which rules are set is a source of even greater power. It is interesting in this context to consider the implications for climate change policies and outcomes of electoral rules in the United States and United Kingdom which appear to have been significantly more vulnerable to gaming and manipulation than electoral systems in Germany and New Zealand.

Changing the underlying paradigms and core goals of a system, from maximising short-term profit to long-term ecological sustainability is also likely to drive far long-lasting transformational change than actions confined to changing isolated targets and feedback loops. Meadows proposes the following checklist of strategies for accelerating a shift in paradigms: sharpen evidence and strengthen understanding of the risks and failures of the old paradigm; visualise and communicate the desirability of alternative futures with as much clarity and power as possible; spend less time focusing on the negative messages of cynics and reactionaries and more time encouraging champions of the new paradigm who are creating living, vivid examples, stories and pictures of new pathways and ways of living.

Meadows also thoughtfully and provocatively concludes with the observation that there is one leverage point even more powerful than paradigm change. 'That is....to stay flexible, to realize that NO paradigm is "true," that everyone, including the one that sweetly shapes your own worldview, is a tremendously limited understanding of an immense and amazing universe that is far beyond human comprehension' [275, p. 164].

Meadows' awareness of the limits to our understanding of the complex unpredictability of self-organising, non-linear feedback systems led her to the realisation that 'we can't control systems or figure them out. But we can dance with them! Before I began to study systems I had learned about dancing with great powers from white water kayaking, from gardening, from playing music, from skiing. All those endeavours require one to stay wide awake, pay close attention, participate flat out, and respond to feedback' [275, p. 165]. Here are some of Meadows's suggestions for learning the art of 'dancing with systems' [275, p. 170].

> *Get the beat – and listen to the wisdom of the system.* Before you disturb the system in any way, watch how it behaves. If it's a piece of music or a white water rapid or a fluctuation in a commodity price, study its beat. If it's a social system, watch it work. Before you charge in to make things better, pay attention to the value of what's already there.

> *Stay humble. Stay a learner.* In a world of complex systems it is not appropriate to charge forward with rigid, undeviating directives....What's appropriate when you're learning is small steps, constant monitoring, and a willingness to change course as you find out more about where it's leading.

Pay attention to what is important, not just what is quantifiable. Go for the good of the whole. Aim to enhance total systems properties, such as creativity, stability, diversity, resilience, and sustainability–whether they are easily measured or not.

Expand time horizons. Expand thought horizons When you're walking along a tricky, curving, unknown, surprising, obstacle-strewn path, you'd be a fool to keep your head down and look just at the next step in front of you. You'd be equally a fool just to peer far ahead and never notice what's immediately under your feet. You need to be watching both the short and the long term–the whole system.

*Expand the boundary of caring. Celebrate complexity....*No part of the human race is separate either from other human beings or from the global ecosystem. The universe is messy. It is nonlinear, turbulent and chaotic. It is dynamic. It self-organizes and evolves. It creates diversity, not uniformity. That's what makes the world interesting, that's what makes it beautiful, and that's what makes it work.

In 2001, in one of her last essays, Meadows asked herself a question central to many challenges explored in this book. 'Can I give my friend, you, myself any honest hope that our world will not fall apart? Does our only possible future consist of watching the disappearance of the polar bear, the whale, the tiger, the elephant, the redwood tree, the coral reef....?' Her response was typically blunt.

> Heck, I don't know. There's only one thing I do know. If we believe that it's effectively over, that we are fatally flawed, that the most greedy and short-sighted among us will always be permitted to rule, that we can never constrain our consumption and destruction, that each of us is too small and helpless to do anything....it's over.
> We are not helpless and there is nothing wrong with us except the strange belief that we are helpless and there's something wrong with us. All we need to do, for the bear and ourselves, is to stop letting that belief paralyze our minds, hearts, and souls. [276]

Shallow and Deep Ecology

Norwegian mountaineer, resistance leader, philosopher and ecological activist Arne Naess first articulated the distinction between 'shallow' and 'deep' ecology in 1973 [277]. In this view, the dominant environmental policy paradigm of 'shallow ecology' remains constrained by an overly anthropocentric view in which the primary goal is continual improvement in human living standards through 'sustainable' exploitation of the Earth's resources.

From this standpoint, 'sustainable' climate action and economic development could be seen as successful if the material living conditions of most human beings continued to improve even if there was a significant decline in the diversity and wellbeing of other species and ecologies. While acknowledging the need for respectful conversations and strategic alliances with a wide range of human rights, social justice and ecological movements, Naess argues that shallow ecology is underpinned by three deeply flawed assumptions [277]. First, that technical innovation and ingenuity can enable human beings to endlessly expand economic consumption even in a world of rapidly escalating climate risks and severely diminished biological diversity. Second that human life is more valuable than the lives and wellbeing of all other species. And third that it is possible and desirable to understand each human individual as a separate entity largely disconnected from other human and non-human life forms.

The theory and practice of deep ecology should therefore, according to Naess, be based on far stronger recognition of the intrinsic value of all life; the necessity of biological and ecological diversity; and the desirability of prioritising 'life quality' over an endless quest for higher material standards of living [278]. Stressing the inherent value of all life forms does not, in this view, necessarily imply a strictly equal ranking of all species. 'In wintertime,' Naess observes 'my cottage receives mice and men as guests, but my obligations are enormously greater toward the human guests than toward the mice' [279, p. 549]. His suggestion that we should also consider broadening the scope of 'non-human life' to include rivers, mountains and oceans as well as animals, plants and human beings continues to resonate strongly with advocates for the ethical and legal 'rights of nature.' This includes the architects and authors of the *2010 Declaration of the Rights of Mother Earth* discussed further in the following chapter and included in Appendix C.

The deep ecology propositions that human beings have no right to reduce the diversity of other life forms except 'to satisfy vital needs' and that the flourishing of non-human life in fact requires a decrease in human population numbers remain highly contentious. The idea that sustaining the richness of non-human life may in fact require a decrease in human population has led some critics to accuse Naess of inciting misanthropic agendas which prioritise the lives of monkeys, bees and plants over human beings. Naess is however careful to clarify the long-term time horizons he has in mind, emphasising that this is an argument for carefully considered, carefully planned population policy implemented with all necessary ethical and democratic safeguards.

The emphasis in deep ecology on the idea that quality of life cannot simply be equated with higher levels of consumption continues to underpin arguments that economic growth may be best framed as a discussion about the quality rather than the quantity of growth. From this standpoint, our wellbeing and enjoyment of life might best be strengthened by having more time to walk on the beach with friends and family rather than owning ever larger television screens. This is, as noted in the earlier discussion on Buddhist economics, exactly the argument espoused by the Bhutanese designers of the *Gross National Happiness* wellbeing measurement framework.

In reflecting on the forces capable of driving transformational change at the necessary speed and scale, Naess points to the distinction originally made by Immanuel Kant between 'moral' and 'beautiful' action. A moral action for Kant is a choice we make based on a sense of duty, even if that choice contradicts our desires and inclinations. A beautiful or joyful action is a choice we make which is consistent with our desires and inclinations. We may therefore be more likely to support transformational change towards more ecological ways of life if we believe our actions will increase our wellbeing and enjoyment rather than feeling a sense of guilt or obligation to follow externally imposed rules. Naess is concerned that 'extensive moralizing within environmentalism has given the public the false impression that we primarily ask them to sacrifice, to show more responsibility, more concern, better morality. As I see it, we need to emphasize the immense variety of sources of joy which are available to people through an increased sensitivity towards the richness and diversity of life, and the landscapes of free nature...' [280, p. 26].

Naess also shares with Donella Meadows the view that advocacy for action in ways consistent with ecologically informed principles needs to be tempered with a healthy dose of humility. 'In the fight for what we believe is right and proper, we must absolutely be rock solid. At the same time, we must always be open to the possibility that we have made a mistake, and be adaptable when new, relevant circumstances arise.'

Naess constantly returns to the challenge of overcoming paralysis and despair. In one conversation, framed as an imaginary interview with himself, he provides the following reflections about the scope and time scale of the challenges unfolding before us.

> Interviewer (I): Are you an optimist or a pessimist?
> Arne Naess (AN): Optimist!
> I: (Astonished) Really?
> AN: Yes, convinced optimist—When it comes to the 22nd Century.
> I: You mean of course the 21st?
> AN: 22nd! The life of the grandchildren of our grandchildren. Are you not interested in the world of your grandchildren!
> I: You mean we can relax because we have a lot of time available to overcome the ecological crisis?
> AN: How terrible, shamefully bad conditions will be in the 21st Century, or how far down we have to start on the way up, depends on what you, you, and others do today and tomorrow. There is not a single day to be lost. We need activism on a high level immediately. [281, p. 432]

Ecologically informed, courageous and decisive action is, according to Naess, the crucial antidote to paralysis and despair. 'The remedy (or psychotherapy) against sadness caused by the world's misery is to do something about it....' [282].

We Are Not the Centre

Tim Morton, described by one recent reviewer as 'the philosopher prophet of the Anthropocene,' holds the Rita Shea Guffey Chair in English at *Rice University* in Houston, Texas [283]. Morton's influence extends well beyond academia to include conversations and collaborations with artist Olafur Elliason, actor Steve Coogan and musicians Pharell Williams and Laurie Anderson. The Icelandic composer Bjork points to one likely source of Morton's growing influence among ecological and climate activists in speaking about her decision to ask him to introduce her 2015 exhibition at New York's *Museum of Modern Art*. 'I guess he is SWERVING the apocalypse angle into hope....' [284].

Laura Copelin, co-curator of Morton's 2018 *Hyperobjects* exhibition at Houston's *Marfa Ballroom Gallery*, wonders if Morton is also trying to 'kick-start an ontological upgrade' in order to help us deal more constructively with climate grief and ecological anxiety. 'To tackle these problems you have to update your notions of art and the world, get past Copernicus, and understand that we are not the centre...' [284].

Readers puzzling over the meaning and relevance of an 'ontological upgrade' may find it useful to understand 'ontology' as the study of how all things, including human beings, exist and relate to each other. Morton's commentary on the need to review and upgrade our view of ontology begins by identifying two very different ways of making sense of the world. One point of view, dominant in Western societies for several thousand years, gives the human species a central, uniquely privileged role at the core of all existence. In this paradigm, an object only really, fully exists because a human being is observing it or interacting with it in some way. A tree therefore is only fully 'real' if a human being can see it or has planted it or cuts it down. We might therefore choose to call this approach to ontology 'human centred' or perhaps 'subject oriented.'

The alternative view, which Morton sometimes refers to as 'Object Oriented Ontology,' starts from the assumption that all objects—a tree falling in a forest; the bacteria in our stomach; the stars in the sky—are all equally 'real' regardless of human perception or action. From this perspective, the vast mesh of objects which make up the universe are interconnected and interdependent in infinitely complex and intricate patterns.

Morton points us to the Buddhist metaphor of *Indra's Net* to help us visualise this way of seeing the world. 'At every connection in this infinite net hangs a magnificently polished and infinitely faceted jewel which reflects in each of its facets all of the facets of every other jewel in the net. Since the net itself, the number of jewels and the facets of every jewel are infinite the number of reflections is infinite as well' [285, p. 40].

All objects, Morton points out, are made up of other objects—like the millions of species making up the ecosystems of the Amazon, the microbes that live inside every human being or the CO_2 molecules that contribute to greenhouse gases. Morton coins the term 'hyperobjects' to describe 'objects,' 'like global warming, that are huge in size and scale and which may develop and evolve over vast periods of time' [286, p. 22].

The observation that 'hyperobjects' are no more or less important than the objects which exist inside them leads us to Morton's slightly counter-intuitive proposition that the whole is actually less than the sum of its parts. A coral reef is an object, as is each of the billions of fish and polyps which create and sustain the reef. The reef cannot exist without the polyps and the polyps cannot exist without the reef. The Earth's climate is comprised of a vast constellation of weather events which are in turn continually and profoundly affected by the changing climate. It is therefore just as misleading to say there is no climate, only weather as it is to focus only on global climate trends without paying close attention to the geographically specific interplay of specific weather events.

Heightened awareness that human beings and the human species are no more, and also no less important than all other beings and species does not, in Morton's view 'mean that we hate humans and want ourselves to go extinct. What it means is seeing how we humans are included in the biosphere as one being among others. This means changing our relationship with the other entities in the universe – whether animal, vegetable or mineral–from one of exploitation through science to one of solidarity in ignorance' [287].

The provocative phrase, 'solidarity in ignorance,' reflects Morton's wariness about the limits of human knowledge and the need for far greater care in the way we throw around terms like 'truth' and 'facts.' We might, therefore, need to be particularly careful in assuming that we can ever fully understand vast 'hyperobjects' like climate change. Following a line of argument informed by quantum physics, Morton asks us to remember that all knowledge about the world is partial. 'Every kind of access – a philosopher thinking about a stone, a scientist smashing a particle, a farmer watering a tree – is profoundly limited and incomplete.'

Morton suggests that the inevitable incompleteness of our knowledge of any object helps us understand the significance of Kant's concept of beauty as a feeling of 'ungraspability.' 'Beauty' he argues 'gives you a fantastic, 'impossible' access to the inaccessible, to the withdrawn, open qualities of things, their mysterious reality' [286, p. 41]. This may help explain the desire which many of us experience to seek out 'beauty' in the form of art and music and poetry and in the natural world in times of anxiety, despair and grief. It also brings to mind to me Ivan Karamazov's celebration of the sustaining quality of blue sky, good soup and 'sticky little leaves' in Dostoevsky's novel, *The Brothers Karamazov*.

> I have a longing for life, and I go on living in spite of logic. Though I may not believe in the order of the universe, yet I love the sticky little leaves as they open in spring. I love the blue sky, I love some people, whom one loves you know sometimes without knowing why.....Here they have brought the soup for you, eat it, it will do you good. It's first-rate soup, they know how to make it here......I love the sticky leaves in spring, the blue sky–that's all it is. It's not a matter of intellect or logic, it's loving with one's inside, with one's stomach [288, p. 208].

Failure to pay sufficient attention to the risks of scientific hubris and the simplistic and selective use of scientific data may also help us make sense of the discomfort many people feel in being 'hit over the head by scientific "factoids" about climate change and ecological risks. When people use factoids, we feel like we are being manipulated by little bits of truth that have been broken off some larger, truer edifice, as if they were small chunks of cake' [286, p. 8].

One of the ironies of constantly evolving debates about the Anthropocene, the awareness that the human species (particularly its wealthiest and most powerful members) is now fundamentally reshaping the Earth's ecology, is the realisation that our species is not, in fact, all-knowing or all-powerful. Indeed, the full extent of the ways in which human lives are enmeshed with other species becomes clearer with every new discovery about the consequences for human beings of the extinction of insects or the loss of biodiversity. We would be wise therefore to be wary of the unintended consequences we might unleash in attempting to solve the 'super wicked' problems of global warming through geo-engineering, AI or emigration to Mars.

Everyday life in the Anthropocene is, Morton notes, coloured for many people by uncomfortable feelings of guilt and shame resulting from the daily avalanche of news about the devastating climatic and ecological consequences of the cars we drive, the holidays we take, the food we eat and the waste we throw away. 'Part of what's so uncomfortable about this is that our individual acts may be statistically and morally insignificant, but when you multiply them millions and billions of times–as they are performed by an entire species–they are a collective act of ecological destruction' [287]. While some people construct finely tuned defence mechanisms to help deny and block this awareness, many others become overwhelmed by grief and guilt. Greater awareness of the partial and contingent nature of human knowledge might therefore help us maintain a healthier, more balanced perspective about the limitations of our individual responsibility.

Paying careful attention to climate emergency risks and drivers as well as to the amount of meat we eat, the number of flights we take and the number of demonstrations we attend are becoming essential preconditions for a responsible way of life on a warming planet. Our capacity to keep paying attention and to sustain responsible, courageous and effective action may however be strengthened if we keep in mind the evidence that just 100 global companies have been responsible for over 70% of global emissions since 1988 [289].

Morton builds on his cautionary advice about the limits of individual knowledge and power to argue that our capacity to sustain effective action in the long emergency of climate change might be enhanced by learning to live a little more 'light-heartedly,' informed by a spirit of 'playful seriousness' [290]. While appearing to trivialise the gravity of the climate crisis may at first glance appear unwise, the theory and practice of 'playful seriousness' and 'serious playfulness' in fact have a long and influential history.

The American educationalist Thomas Dewey wrote at length about the value of finding the right mix of playfulness and seriousness in maximising learning and creative outcomes. Martial arts teacher and film director Bruce Lee famously observed 'turn your sparring into play but play seriously.' US Olympic fencing champion, Charles Selberg, suggests that 'fencing is the epitome of serious playfulness. The intensity of concentration, the pouring out of yourself experienced in sparring resembles the seriousness of a child at play. Play is an attitude, not an activity; it is a freedom and creativity that comes from the courage of complete commitment' [291, p. 10].

I suspect Morton is drawing on some of these ideas in making the connection between playfulness and comedy: 'How do we go from tragedy mode to comedy mode? Comedy doesn't mean this is funny. Comedy means you allow all the emotions, not just fear and pity, to coexist….I think comedy is deeper than tragedy. When you can laugh, you can cry. This is grief work' [292]. This observation might also help make sense of the tendency which many climate scientists and activists refer to of maintaining a strong sense of the absurd and of black humour as part of their repertoire for dealing with confronting and threatening realities.

Where Should We Land?

French anthropologist and philosopher, Bruno Latour was born into a wine-growing family near the town of Beaune in Burgundy in 1947. While family background and place of birth can help us locate the work of many writers, the rich, brocaided patterning of Burgundy's environmental, agricultural and industrial landscapes provide important clues in understanding Latour's response to his central questions: How can we find our way; where should we land and how should we live in the increasingly alien terrain and era of the Anthropocene? Indeed, as one recent commentary on Latour's work notes, 'the dense fabric of grapevines, terroirs, rhizobacteria, fermentation processes, the vintner's art, glass industry, and haulage firms, which is so typical for the Beaune context, represent the basic model for Latour's recurrent theme that natures, humans, things, and technologies belong together' [293, p. 10].

Latour's diverse and influential contributions to debates about the interdependence between human beings and nature; science and technology; knowledge, modernity and power span many hundreds of publications. The following brief reflections focus primarily on key themes explored in his 2018 book, *Down to Earth: Politics in the New Climate Regime* [294].

Latour's starting point is the idea that contestation over the causes and consequences of climate change has, over the last forty years, become the central driver of social and political change. 'Without the idea that we have entered into a New Climatic Regime, we cannot understand the explosion of inequalities, the scope of deregulation, the critique of globalization, or, most importantly, the panicky desire to return to the old protections of the nation-state' [294, p. 2].

The roots of our current crisis can be found, Latour suggests, in the 1970s and 1980s as corporate boards and CEOs became increasingly concerned about the risks of unconstrained expansion in resource consumption and rapidly rising CO_2 emissions. Publicly, the dominant 1980s corporate discourse remained fiercely critical of emission reduction policies and arguments about the unsustainability of endlessly accelerating economic growth. Privately, the view behind closed board room doors was often very different as recent revelations about the full extent of the knowledge of oil company executives about climate change trends and risks have shown.

As far back as 1977, Exxon's senior scientist James Black advised Exxon's Management Committee for example that 'there is general scientific agreement that the most likely manner in which mankind is influencing the global climate is through carbon dioxide release from the burning of fossil fuels' [295]. This advice does not appear to have significantly altered the company's commitment to spend millions of dollars on think tanks focused on undermining trust in climate science.

Latour's contention is that powerful sections of the corporate world had already, by the 1980s, come to accept much of the key evidence about climate risk as well as the problematic nature of the goal of infinite growth on a finite planet. Many corporations however drew very different conclusions to environmental and climate activists about the implications of this view.

> The elites have been so thoroughly convinced that there would be no future life for everyone that they have decided to get rid of all the burdens of solidarity as fast as possible–hence deregulation; they have decided that a sort of gilded fortress would have to be built for those (a small percentage) who would be able to make it through –hence the explosion of inequalities; and they have decided that, to conceal the crass selfishness of such a flight out of the shared world, they would have to reject absolutely the threat at the origin of this headlong flight—hence the denial of climate change. [294, p. 18]

Latour illustrates his analysis with an alternative version of the familiar analogy of the sinking of *The Titanic*. In this version, the first-class passengers fully understand the terrifying implications of their briefing from the captain about the imminent, inevitable collision with the iceberg. They decide that the only way to ensure that all first-class ticket holders have places on the lifeboats is to deliberately conceal the catastrophic reality of the impending disaster from all other passengers; undermine trust in the crew and spread doubt and confusion about the reliability of messages being received from scientists about the reality of icebergs; encourage the band to keep playing lively tunes to divert attention and cover up the sound of shrieking metal and rushing water; pit different groups of passengers, particularly those from different ethnic or racial backgrounds against each other......

Fast forward, Latour suggests to the weeks following the *2015 Paris COP* where initial elation about the *Paris Climate Agreement* is rapidly overtaken by foreboding about the limited capacity of national governments to implement policies required to keep global warming below 1.5 or even 2 degrees C. 'On that December day all the signatory countries, even as they were applauding the success of the improbable agreement, realized with alarm that, if they all went ahead according to the terms of their respective modernization plans, there would be no planet compatible with their hopes for development' [294, p. 5].

This alarm was further heightened less than twelve months later by the election, in 2016 of Donald Trump as President of the United States. By June 1, 2017, the date on which the United States withdrew from the Paris Agreement, the full extent of the danger had become even clearer. 'From the 1980s on, the ruling classes stopped purporting to lead and began instead to shelter themselves from the world. We are experiencing all the consequences of this flight, of which Donald Trump is merely a symbol, one among others. The absence of a common world we can share is driving us crazy.'

> To resist this loss of a common orientation, we shall have to come down to Earth; we shall have to land somewhere. So, we shall have to learn how to get our bearings, how to orient ourselves. And to do this we need something like a map of the positions imposed by the new landscape within which not only the affects of public life but also its stakes are being redefined. [294, p. 2]

Foreboding and anxiety about the absence of a common world have been further compounded by widespread uncertainty about the direction we are heading and the extent to which we should be prioritising globalising or localising goals and strategies. Latour illustrates the seriousness of our dilemma in the following way. 'Let us call it a conflict between modern humans who believe they are alone in the Holocene, in flight toward the Global or in exodus toward the Local, and the terrestrials who know they are in the Anthropocene and who seek to cohabit with other terrestrials under the authority of a power that as yet lacks any political institution' [294, p. 90].

The rich and powerful have, over the last 50 years, been very successful in convincing many people of the desirability and inevitability of endlessly accelerating technological progress towards a globalised nirvana. In addition to the tantalising prospect of ever-increasing material prosperity, some advocates of globalising capitalism have been keen to celebrate the attractions of more cosmopolitan ways of life, more respectful and inclusive of multiple cultural and political viewpoints. The price paid for this globalising cosmopolitanism has sometimes been the loss of connection to local places, economies and cultures. The even deeper problem is that it has now become clear to those on the losing end of globalisation that the promise of a prosperous, cosmopolitan future for all is increasingly illusory.

Many people therefore have become attracted to the possibility of returning to the comforting security and clear identity of local places and economies. Latour notes that the attraction of the local lies in 'the importance of history, of the basic right to feel safe at home, and to cultivate care and attachments to the soil'. Localist ideas and strategies have the potential to open pathways to more grounded, convivial and interconnected ways of life. They can also be 'easily coopted by nationalism and rigid concepts of identity and borders' leading to the walled cities and nations of Johnson and Trump and Bolsonaro; of Brexit and *Stop the Boats* and *Make America Great Again*.

To illustrate the problem that we can't go forward and we can't go back, Latour asks us to imagine we are on an aeroplane flying at great speed towards the future. Halfway through our flight, we listen with growing alarm as the pilot informs us that our initial destination is no longer available as the landing strip and surrounding countryside have been flooded and destroyed by wild storms and surging waves. Nor is it possible for us to return to our place of departure which has now been overwhelmed by fire and smoke. One of the passengers from the first-class cabin, introducing himself as Donald Trump, leaps up and points wildly out the window calling on the pilot to change direction for Mars. Some passengers applaud. Many others remain deeply unconvinced. But where else can we land if we can't go forward and we can't go back?

> Now if there is no planet, no Earth, no soil, no territory to house the Globe of globalization toward which all these countries claim to be headed, then there is no longer an assured 'homeland,' as it were, for anyone. Each of us thus faces the following question: Do we continue to nourish dreams of escaping, or do we start seeking a territory that we and our children can inhabit? Either we deny the existence of the problem, or else we look for a place to land. [294, p. 5]

Latour coins the term 'terrestrial' to describe and locate the places in which we might seek to safely land. Unlike the destination of a 'global' future, safe arrival on a 'terrestrial' landing ground calls for clearer understanding that human beings do not comfortably exist above and beyond the Earth. In this view, we are generally likely to be most content when grounded and embedded in the Earth's soil and systems and relationships. Unlike a narrowly 'local' future, 'designed to differentiate itself by closing itself off, the Terrestrial is designed to differentiate itself by opening itself up. The Terrestrial is bound to the Earth and to land, but it is also a way of worlding, in that it aligns with no borders, transcends all identities.'

Latour is well aware that 'the Achilles' heel of any text that purports to channel political affects toward new stakes is that the reader can justifiably ask: "All that is well and good. The hypothesis may be attractive, though it still waits to be proved, but what are we to do with it, practically speaking, and what does it change for me? Do I have to take up permaculture, lead demonstrations, march on the Winter Palace, follow the teachings of St. Francis, become a hacker, organize neighbourhood get-togethers, reinvent witches' rites, invest in artificial photosynthesis, or would you rather I track wolves?"' [294, pp. 90–91]

In reflecting on Latour's provocative wide-ranging list, I am aware of many people I admire and respect who are testing and exploring some or all of these possibilities. In doing so, they also often keep returning to Latour's central question. Where can we land? How, on a rapidly heating planet, do we imagine and create ways of living which avoid the equally unhelpful dead ends of technocratic hubris and parochial tribalism?

Homecoming to the Earth

The Australian feminist and ecological author and activist, Val Plumwood, provides a number of helpful compass settings for locating promising sites to land on a strange and unfamiliar planet. Plumwood's argument, explored with clarity and precision in her 1993 work, *Feminism and the Mastery of Nature*, proceeds in the following way [296].

The strongest driver of the climatic and ecological crises we now face is the dominant view that some human beings exist in a realm of mastery separate from and far above other human beings and the rest of nature.

This view is underpinned by the conveniently self-serving claim that a small minority of powerful and privileged individuals are uniquely endowed with reason and logic, narrowly defined as the power of the human mind to analyse, control and colonise the world.

Reason and logic are honoured and celebrated in this narrative as the active, energetic drivers of prosperity and progress. The realm of nature, often framed by the powerful as encompassing women and colonised peoples as well as other species, is defined as passive, inferior and subordinate. If mastery over the Earth and other beings depends on the power of reason, then reason is primarily the domain of the master.

While noting that the strongest advocates for and beneficiaries of this view do indeed tend to be men, Plumwood also points us to the work of Indian feminist and biodiversity scholar and activist Vandana Shiva in foregrounding the importance of class and race as well as gender as sources and drivers of inequalities of power and wealth. 'It is' she notes 'not only women's labour which traditionally gets subsumed.....into nature, but the labour of colonised non-western, non-white people also' [296, p. 4].

Plumwood outlines a series of interconnected steps in the evolution of the dominant narrative 'in which reason progressively divides, devalues and denies the colonised other which is nature.' Plato, she argues, sets the scene by portraying non-human nature as 'a chaotic and formless void' and by welcoming to the stage an oligarchy of philosopher kings divinely ordained to rule over 'the lower orders.....animals, slaves, barbarians and women' [296, p. 25].

Enlightenment philosophers like Descartes and Locke build on this foundation by defining non-human nature as 'terra nullius, uninhabited by mind....and totally available for annexation.....a sphere easily moulded to the ends of a reason conceived as without limits' [296, p. 4]. The doctrines of 'reason without limits' and of 'terra nullius' went on to provide European colonisers with powerful ethical and legal justifications for the invasion of Australia, Asia, Africa and the Americas; the conquest and enslavement of their Indigenous inhabitants; the destruction of their forests and the slaughter of vast numbers of buffalo, seals and whales.

In the concluding act, 'the age of devouring,' 'subordinate and colonised human beings lose all intrinsic value and become an inert, passive resource freely available to be commodified and consumed....Reason systematically devours the other of nature....the operations of the Rational Economy become as destructive of the sociosphere as they are of the biosphere' [296, p. 194]

Here, Plumwood argues, is the dark place in which we have now arrived—a world in which rising numbers of human beings and other species are devoured by the fires and overwhelmed by the storms of the climate emergency sweeping over us and a world too in which rich and powerful masters of the universe continue to champion technological solutions such as carbon capture and storage and geoengineering in arguing against the need for significant reductions in consumption.

Plumwood's framing of the ecologically orientated feminism which she proposes as the most promising candidate for an alternative political paradigm carefully avoids romanticised assumptions about the inherently nurturing ecological awareness of women. While agreeing that the world we now live in does indeed increasingly resemble a feminist dystopia of misogynistic violence, burning forests and mass extinctions, she is wary of a narrative which portrays all women as ecological angels and all men as ecological vandals.

> Accounts of a generalised 'patriarchy' as the villain behind the ecological crisis implicitly assume that western culture is human culture. But the gendered character of nature/culture dualism, and of the whole web of other dualisms interconnected with it, is not a feature of human thought or culture per se, and does not relate the universal man to the universal woman; it is specifically a feature of western thought. [296, p. 11]

This more nuanced understanding of the interconnected sources and drivers of ecological crises and deepening inequalities of class, race and gender leads Plumwood to advocate for a rationality focused less on mastery and power and more on 'mutually sustaining relationships between humans and the Earth....the rationality of the mutual self, the self which can take joy in the flourishing of others, which can acknowledge kinship but also feast on the other's resistance and grow strong on their difference' [296, p. 196]

Identifying and creating the landing zones and ways of life required for human beings to survive and thrive in the harsher, hotter world in which we now find ourselves will, in Plumwood's view, require us to learn to tell new stories about the values, virtues and practices of mutual trust and mutual respect; co-operation, care and reciprocity.

> Much inspiration for new, less destructive guiding stories can be drawn from sources other than the master, from subordinated and ignored parts of western culture, such as women's stories of care. Those of us from the master culture who lack imagination can gain new ideas from a study, undertaken in humility and sympathy, of the sustaining stories of the cultures we have cast as outside reason.
>
> The master culture must now make its long-overdue homecoming to the Earth. This is no longer simply a matter of justice, but now also a matter of survival. [296, p. 6]

Ecological awareness, insights and experiences provide many timely and inspiring starting points for designing and implementing strategies of climate justice and survival. Deepening our awareness of ecological fragility and beauty, complexity and power can also, Vandana Shiva argues, open our eyes to pathways and landscapes of courage, hope and joy.

> My joy comes from the awareness that I'm part of this amazing Earth. I'm part of this amazing universe. And then the next step in the build-up of joy comes from flowing with the paths that are laid out for us. You can call them ecological paths. We can look at the collapse and despair and say 'Oh my god, we are going over the precipice,' or you can take that one seed and plant it, and ask everyone around you to grow their own seeds of change, seeds of joy, seeds of freedom, seeds of hope. [297]

10

How the Light Gets In: Imagining and Creating Just and Resilient Zero-Carbon Worlds

I began this book by speaking about the importance of facing the consequences of climate emergency and ecological tipping points with eyes wide open. A fully frank and honest assessment of this evidence provides a sharp, clear warning about the urgency of action and of the likelihood that even the most decisive action will drive current and future generations on a journey into increasingly harsh and unfamiliar lands. My search for maps and guides to help us navigate this strange new world often leads me to the wise advice of Rob Hopkins, founder of the Transition Towns movement: 'Imagination is the only thing we have that is – or could be – radical enough to get us through, provided it is accompanied, of course, by bravery, and by action' [298, p. 13].

Imagination, closely accompanied by bravery and action, is critically important for three reasons. First, imagination understood as the 'ability to look at things as if they could be otherwise' can provide disruptive counter-narratives to the toxic, paralysing myth that 'there is no alternative' to currently dominant and profoundly unsustainable social, economic and political paradigms [298]. Second, the capacity to visualise and communicate alternative social, technological and ecological imaginaries is an essential foundation for the transformational changes required to build just and resilient zero-carbon societies. Third, alongside emergency speed emission reduction, we also now need to prioritise imagining and creating regenerative and replenishing ways of living which can help us traverse the alien landscapes and ecologies which lie before us.

Other Worlds Are Possible

The dominant paradigms underpinning the political and economic systems responsible for a significant proportion of GHG emissions over the last 50 years are closely aligned with the neoliberal framing first articulated by former British Prime Minister Margaret Thatcher: 'There is no such thing as society' and 'There Is No Alternative.' The narrative of competitive and acquisitive individualism as the essential, inevitable pinnacle of human evolution has also been an influential driver of the forces transforming globally interconnected citizens and communities into alienated workers and atomised consumers.

The vast resources invested in advertising and marketing combined with increasingly sophisticated manipulation of social media platforms and algorithms have also played a key role in accelerating unconstrained and unsustainable consumption. The deeper and more corrosive impacts of these investments are also visible in the destruction of trust in public institutions and the rapid expansion of commercialised and transactional interpersonal relationships. The post-truth politics of Trump and Johnson; Bannon and Murdoch have taken these dark arts to another level.

The English critical social theorist, Mark Fisher has coined the term 'capitalist realism' to describe 'the widespread sense that not only is capitalism the only viable political and economic system, but also that it is now impossible even to imagine a coherent alternative to it' [299, p. 2]. The first task of emancipatory politics, Fisher argues, is therefore to reveal 'what is presented as necessary and inevitable to be a mere contingency, just as it must make what was previously deemed to be impossible seem attainable' [299, p. 17].

Dutch urban theorists Maarten Hajer and Wytske Versteeg identify four strategies for fracturing dominant assumptions about the impossibility of visualising and creating a post–fossil-fuel world [300]. The first strategy, the discourse of limits, relies on broadening awareness about the dire consequences of continuing to consume resources at ever-increasing levels. This is the logic informing and framing the messaging of 'climate emergency' and 'limits to growth.' While many activists find these messages compelling, it remains unclear whether concerns about future impacts can, on their own, overcome the tendency for many of us to focus our gaze on more immediate and tangible threats.

The second approach, problematising the present, aims to strengthen understanding of the causes and drivers of catastrophic events such as uncontrollable fires and unbreathable air. Key challenges here include joining the dots between cause and effect and overcoming the temptation to prioritise short-term disaster management over longer-term preventative action. The third and fourth possibilities, demonstrating the feasibility and desirability of alternative, more sustainable ways of life and refusing to accept the taken for granted legitimacy of dominant sources of power and authority, both highlight the crucial role of skilful story telling. This is the critical task which George Monbiot points us to in emphasising the power of accessible and compelling narratives which fully 'address the gravity of our situation yet which also create longing, which evoke a deep and rich sense of the wonders we can still create' [298, p. 126].

> You cannot take away someone's story without giving them a new one. It is not enough to challenge an old narrative, however outdated and discredited it may be. Change happens only when you replace it with another. When we develop the right story, and learn how to tell it, it will infect the minds of people across the political spectrum. Those who tell the stories run the world. [301]

The desire to visualise and enact socio-technical imaginaries and pathways better aligned with the goal of keeping global warming below 1.5 degrees is also reigniting interest in utopian narratives and methodologies. Historian of utopian ideas, Ruth Levitas suggests 'the utopian approach allows us not only to imagine what an alternative society could look like, but enables us to imagine what it might feel like to inhabit it' [302, p. 3].

Utopian projects, Levitas argues, are best understood as methodologies for provoking and informing conversations about the necessity and possibility of radical social change rather than as a set of specific blueprints. Utopian narratives and images can sharpen awareness of the gap between current realities and alternative possibilities as well as providing glimpses and sketches of prefigurative experiments through which we can begin to explore and test other, more desirable ways of living and working.

Science fiction writer and social theorist China Miéville employs the metaphor of the Rorschachs inkblot psychological test to describe and celebrate the disruptive potential of utopian imaginaries. 'We pour our concerns and ideas out, and then in dreaming we fold the paper to open it again and reveal startling patterns. We may pour with a degree of intent, but what we make is beyond precise planning' [303]. Miéville cautions us however to keep a sharp eye on the dangers of naïve utopianism. 'Utopias are necessary. But not only are they insufficient: they can, in some iterations, be part of the ideology of the system, the bad totality that organises us, warms the skies, and condemns millions to peonage on garbage scree.' Utopian fantasies of escaping a burning Earth by colonising Mars are unlikely to provide refuge for more than a tiny proportion of the human population. Our hopes and dreams, Miéville argues, must be informed by a sharp focus on who exactly stands to win and lose. 'We must learn to hope with teeth' [303].

Socialist author and activist Erik Olin Wright's overview of strategies which can help us to 'hope with teeth' focuses our attention on the work of building 'real utopias, grounded in the real potentials of humanity, utopian destinations that have accessible way stations' [304, p. 6]. Wright's discussion of the work required to imagine and create 'real utopias' emphasises the importance of plausible and coherent theories of change. Fantastic visions of castles in the sky are worse than useless if they remain unattainable fantasies, further confirming the inevitability of existing paradigms and the futility of action. Wright suggests that a strong and useful theory of transformational change needs to address four key questions.

What forces preserve and reproduce dominant institutions and relationships, even when there is clear evidence that they are causing significant, widespread harms? Direct coercion and repression are often only the most visible component of such forces. Other equally powerful drivers include the 'rules of the game' which protect and reinforce inequalities of wealth and power; the dependence of individuals and communities on existing economic systems to meet immediate needs; and the multifaceted ways in which media and advertising amplify and entrench dominant cultural norms and values. Importantly, Wright concludes, the most powerful of these norms are beliefs about what is possible. 'People can have many complaints about the social world and know that it generates significant harms to themselves and others and yet still believe that such harms are inevitable, that there are no other real possibilities that would make things significantly better and that thus there is little point in struggling to change things, particularly since such struggles involve significant costs' [304, p. 287].

What limits and contradictions have the greatest potential to disrupt or overturn the forces preserving and reproducing dominant and seemingly invulnerable institutions and relationships? Game-changing disruptive forces include trade-offs and conflicts between individual and organisations with differing interests and alliances and the limited capacity of even the most powerful actors to predict and control complex, rapidly evolving socioeconomic, political and ecological developments— pandemics for example, or the catastrophic consequences of increasingly frequent, increasingly severe fires, storms and floods.

What are the underlying dynamics and trajectories of unintended change? Here Wright stresses the opportunities as well as risks of sudden social, technological, economic and ecological shocks and crises. This is the logic informing neoliberal economist Milton Friedman's famously influential advice that 'only a crisis – actual or perceived –produces real change. When that crisis occurs, the actions that are taken depend on the ideas that are lying around…. Our basic function is to develop alternatives to existing policies, to keep them alive and available until the politically impossible becomes the politically inevitable' [305, p. xiv]. While Friedman of course had right-wing, libertarian goals in mind, the same ideas have been creatively repurposed by climate activist Naomi

Klein in her 2007 book *Shock Doctrine* to emphasise their relevance for many other varieties of transformational politics [306].

What are the most effective and strategic ways for collective actors to deliberately intervene to transform the system? Here Wright draws on historical evidence of the importance of well-informed judgement about the time and place for triggering decisive action, bearing in mind Martin Luther King's reflection that, in the end, 'the arc of history bends towards justice.' Up until the beginning of 2020, it seemed that the toughest of all challenges facing climate activists was the difficulty of imagining circumstances powerful and disruptive enough to bend the arc of history and transform the global economy at the necessary speed and scale.

And then COVID-19 arrived.

Imagining the Bridge: Disruptive Transformation in a Post-pandemic World

In December 2019, the Chinese government announced that it had identified the first case of coronavirus disease (COVID-19) in Wuhan. By May 2021, over 150 million people in almost every corner of the world had been diagnosed with COVID-19. Over 3 million people had died. The global economy was heading for its largest contraction since the Great Depression. Hundreds of millions of people lost their jobs. Most countries closed their borders. Most planes were grounded. Most children were educated at home. Some countries began to treat their health, aged care and child care services with a little more respect. Others introduced various forms of basic income protection. A wide range of Orwellian tracing and tracking technologies were fast-tracked and deployed. Sudden, rapid, global disruptive social, political and economic transformation was, it turned out entirely possible.

One of the many cautionary lessons from the COVID-19 experience has been the need for care in predicting the ways in which the impacts and consequences of the pandemic will evolve. At the time of writing, in early 2021 it does however seem that the impact of the disease and the scale of the response have triggered some reconsideration of political and economic priorities. In April 2020, Dr. Elizabeth Sawin, director of the Washington based *Climate Interactive* research institute, offered, for example, the following reflections.

> COVID-19 is many things. A disease. A complex system. A measure of our interconnection. A reason to pause. A threat to our safety. A discontinuity with the future we thought we had. A threat multiplier. An inequity revealer. An economy disrupter. But I'm starting to think it is, in being all of these things, most of all a highly polished mirror, or maybe a broom that sweeps away illusions.
>
> One of our choices in these times is whether we will choose to stare in the mirror, or not. Will we see what is revealed or will we look away and devote mental energy instead to distraction and propping up comforting illusions? Looking in the mirror is work. So is building up the illusions of normalcy. Both efforts might leave us exhausted at the end of a day. But, it seems to me, only one opens the door to a possibly better future. [307]

The greatest challenge for advocates for emergency speed climate action is, science fiction author Kim Stanley Robinson noted in 2011, 'that we can't imagine how we might get there. We can't imagine the bridge over the Great Trench, given the world we're in, and the massively entrenched power of the institutions that shape our lives....the bridge itself is what we can't imagine' [308, p. 19] Perhaps, Robinson wondered in May 2020, 'the virus is rewriting our imaginations. What felt impossible has become thinkable' [309]. Perhaps the pandemic will open the minds of many more citizens and governments to scientific evidence about the action required to address other existential risks like melting ice caps, ocean acidification and mass extinctions.

It might also be possible that learning from societies which have responded most successfully to the pandemic will strengthen understanding that 'society is not only real; it's fundamental. We can't live without it. And now we're beginning to understand that this "we" includes many other creatures and societies in our biosphere and even in ourselves' [309]. Or perhaps our memories will be too short, the desire for immediate reward too great and the dominant structures of corporate wealth and power too strong. Time will tell.

Creating a Just and Resilient Zero-Carbon Future

Christiana Figueres, former Executive Secretary of the *United Nations Framework Commission on Climate Change* and coauthor of *The Future We Choose: Surviving the Climate Crisis,* is well placed to understand the challenges involved in emergency speed bridge building. Figueres outlines the stark choices we now face in the following way.

> If we do not halve our emissions by 2030, we are highly unlikely to be able to halve emissions every decade until we reach net-zero by 2050. That is our final limit. We cannot exceed it. The planet will survive, in changed form no doubt, but it will survive. In deciding what kind of world we and future generations will live in, we don't have many options; we have in fact only two.... [310, p. 16]

The first option, 'the world we are creating,' is the dystopian nightmare of searing heatwaves and raging storms; burning forests and bleaching coral reefs; melting ice and rising tides; thirst and famine; refugees, disease and war. The key features of 'the world we must create,' a zero-carbon world in which we still have some chance of keeping global warming below 1.5 or 2 degrees, are well summarised by young Mexican climate activist, Xiye Bastida [311].

> The biggest tree planting campaign in history is sucking billions of tonnes of carbon out of the air and forests and Indigenous lands are protected....Streets are pedestrian and kid friendly. Food growing on roof tops and in car parks.....There are solar panels on every roof across the globe....clean interconnected energy lights every home. Every clinic every school....Public transport everywhere is electric dependable and free....Farming is all regenerative which means healthy soil and better food. And we don't eat much meat. Can you imagine what we could do with the third of the world's crop land currently used to grow animal feed? Fields of seaweed miles long grown in oceans across the planet. They draw down billions of tonnes of carbon.....

Bastida also shares Figueres' passionate advocacy for the transformational power of creativity, imagination and solidarity, underpinned and sustained by a spirit of tough minded, stubborn optimism.

> Viewing our reality with optimism means recognising that another future is possible, not promised. Optimism is about having steadfast confidence in our ability to solve big challenges. It is about making the choice to tenaciously work to make the current reality better....Optimism is not soft, it is gritty. Every day brings dark news, and no end of people tell us that the world is going to hell. To take the low road is to succumb. To take the high road is to remain constant in the face of uncertainty. [311]

The key features of the zero-carbon pathway visualised by writers like Bastida and Figueres align closely with the most rigorous scientific evidence and advice. The 2017 article, *A Roadmap for Rapid Decarbonisation* written and published in the journal *Science* by six of the world's most eminent climate scientists, provides a concise, science-based plan for peaking emissions by 2020 and then halving gross CO_2 emissions every decade until 2050. [54]

Initial crucial decisions and actions outlined in this plan include globally integrated carbon tax and emission trading schemes, raising the global carbon price from $50 to over $400 per metric ton by midcentury; eliminating fossil fuel subsidies and ending all investment in new unabated coal-based energy projects. By 2030, coal-based energy systems would be almost completely replaced by renewables, underpinned by huge investment in energy efficiency, energy storage and super smart power grids. Electric vehicles and fully electrified public transport would be well on track to replace combustion engine cars. By 2030, we would also see initial results from a massive, worldwide program of tree planting and reforestation combined with rapid progress towards the creation of sustainable plant-based global food and agriculture systems.

By 2040, oil would be exiting the global energy mix. Aircraft fuel would have become entirely carbon neutral, utilising synthesised fuels, bio-methane and hydrogen. All building construction would also have become carbon-neutral or carbon-negative employing emissions-free concrete and steel along with zero or negative-emissions substances such as wood, stone and carbon fibre. By 2050, the world would have reached net-zero CO_2 emissions, with a global economy powered by carbon-free energy and fed by carbon-sequestering sustainable agriculture.

The likelihood that achievement of a zero-carbon economy by 2050 will probably still lead us on a path to well over 2 degrees of global warming still leaves us however with several tough and enduring questions. How do we further accelerate just and sustainable emissions reductions strategies while also visualising and implementing policies and practices of resilience, replenishment and regeneration?

Justice and Care for All Creation

Climate justice, the proposition that the burdens and benefits of climate change and its resolution should be shared equitably and fairly, is now widely understood as an essential ethical and strategic precondition for accelerating economic de-carbonisation. Oxford University Professor philosopher Henry Shue suggests that the climate justice imperative requires us to address three key challenges. 'What is a fair allocation of the costs of avoiding however much additional warming can and reasonable should be avoided? What is a fair allocation of the costs of social adjustments to however much warming is not avoided? How should the responsibility for solving the problem of global warming be divided in light of the background inequalities in wealth and power that are the present bitter fruit of centuries of colonialism, imperialism, inequality; unequal development, war, greed [and] stupidity?' [312, p. 127 and 128].

The 2002 *Bali Principles of Climate Justice* and the 2010 *Universal Declaration of Rights of Mother Earth* (see Attachments B and C) provide two useful starting points for constructing and articulating a comprehensive response to these climate justice challenges [313, 314] The assumptions and insights underpinning both these documents are informed by growing awareness that broader principles of respect and responsibility; care and compassion may be more helpful than the narrower framing of 'rights' and 'justice' in guiding the complex interplay of intertwined and reciprocal relationships between human beings and other species [315].

The *Bali Principles of Climate Justice* developed through a collaborative process involving a broad range of environmental and Indigenous social justice organisations begin by focusing our attention on the right of all human beings, including the poorest and most vulnerable women, rural and indigenous peoples in current and future generations 'to be free from climate change, its related impacts and other forms of ecological destruction'; to have unimpeded access to 'clean air, land, water, food and healthy ecosystems'; and 'to participate effectively at every level of decision-making.'

The *Bali Principles* also include a strong emphasis on the sources of climate injustice....'the role of transnational corporations in shaping unsustainable production and consumption patterns and lifestyles'....and the actions required to ensure that the 'victims of climate change and associated injustices....receive full compensation, restoration, and reparation for loss of land, livelihood and other damages.'

The Declaration of Rights of Mother Earth was a key outcome of the *World People's Conference on Climate Change and the Rights of Mother Earth* held in Cochabamba, Bolivia, in April 2010. While the *Bali Principles* begin by 'affirming the sacredness of Mother Earth, ecological unity and the interdependence of all species,' the articulation of this principle in *The Declaration* is informed by the even more radical premise that 'Mother Earth is a living being...a unique, indivisible, self-regulating community of interrelated beings that sustains, contains and reproduces all beings.'

The Declaration goes on to argue that 'Mother Earth and all beings of which she is composed' have a wide range of inherent rights including the right to 'maintain its identity and integrity as a distinct, self-regulating and interrelated being; the right to water, clean air and integral health; the right to be free from contamination, pollution and toxic or radioactive waste; the right to a place and to play its role in Mother Earth for her harmonious functioning and the right to wellbeing and to live free from torture or cruel treatment by human beings.'

Creating a Practice of Resilience, Replenishment and Regeneration

Alongside emergency speed emissions reduction, we also now need to urgently prioritise imagining and creating the regenerative and replenishing ways of life which can help us navigate the alien landscapes and ecologies which lie before us. Christiana Figueres usefully outlines the scope and scale of these regenerative action challenges.

> Our first responsibility is to notice how and when we are depleted and to support ourselves. Our second responsibility is to reaffirm and strengthen the regenerative capacity we already display with family and friends. But we cannot stop there. Our third responsibility is to engage those beyond our innermost circle and, indeed, nature itself.
>
> We will not recover everything. Many species are already extinct and will not return, and some ecosystems may already be damaged beyond their resilience threshold. But fortunately we still have a relatively hardy natural environment that responds positively to our care and caring. Well-intentioned and well-planned regenerative practices will restore our ecosystems, perhaps not to their former state but to a new state of regained health with enhanced resilience. [310, p. 63]

David Wahl, author of *Designing Regenerative Cultures,* suggests that the defining characteristics of regenerative cultures include: valuing and prioritising the principles and practices of interdependence, diversity and reciprocity and of transformative adaptation and resilience [316]. Transformative adaptation and resilience, Wahl argues, extends well beyond the ability to 'bounce back' and sustain existing systems in response to sudden catastrophic disasters. Societies and cultures which thrive and flourish over many generations need to constantly reinvest in the knowledge, skills and values required to make well-informed, well-considered judgements about potential risks and to make swift, effective and ethical decisions in the face of unexpected shocks.

Awareness continues to grow of the importance of regenerative food production and consumption, farming and land management as essential components of the work required to reverse global warming, strengthen climate justice and restore biodiversity [317]. So too, Vandana Shiva argues, does understanding of the importance of regenerative principles and practices in our work to construct and sustain more hopeful and courageous ways of life.

> Hope is something we cultivate in our daily consciousness through our daily actions. So everyone who is feeling a little hopelessness needs to get engaged with turning bread into a sacrament at every meal....If each of us not just became conscious in our eating, but became engaged in the political and social activism, to recognize that in our daily bread is an interconnectedness of care for the Earth and care for the community, in it is the link between ecological action as well as social justice action and health action and political action... it's all in that food that we eat two to three times a day. [297]

Donna Haraway, author of *Staying With the Trouble,* adds the following reflections on practices of replenishment and regeneration which can help us rise above two equally flawed perspectives on life in deeply troubled times: wildly over-optimistic faith in technofixes or self-fulfilling prophecies of indifference and despair. 'We – all of us on Terra – live in disturbing times, mixed-up times, troubling and turbid times....Staying with the trouble requires learning to be truly present, not as a vanishing pivot between awful or edenic pasts and apocalyptic or salvific futures, but as mortal critters entwined in myriad unfinished configurations of places, times, matters, meanings' [318, p. 1]

When asked which writers I find most helpful in thinking about wisdom and meaning in the long emergency, three books I often turn to are Rebecca Solnit's *A Paradise Built in Hell*, NK Jemisin's *Broken Earth* and Kim Stanley Robinson's *Ministry for the Future*. Solnit's eloquent and insightful guide to the diverse ways in which individuals and communities respond to emergencies and disasters begins by noting the derivation of these words. 'Emergency.....from emerge, to rise out of.....a separation from the familiar, a sudden emergence into a new atmosphere.....Disaster....from the Latin compound of dis or away, without and 'astro' star or planet....literally without a star.....' [319, p. 306]. The long emergency, the journey on which we are now embarking, might therefore usefully be imagined as the experience of waking in a hotter, wetter world in which the sky is full of strange and unfamiliar stars.

Reflecting on the experience of catastrophies like the 1906 San Francisco Earthquake, Hurricane Katrina in New Orleans and 9/11 in New York lead Solnit to the view that most people rise to the challenge of sudden, overwhelming disasters with bravery and kindness; solidarity and compassion. 'Two things matter most about these ephemeral moments. First they demonstrate what is possible or perhaps more accurately latent: the resilience and generosity of those around us and their ability to improvise another kind of society. Second they demonstrate how deeply most of us desire connection, participation, altruism and purposefulness' [319, p. 305].

Solnit is well aware however that the increasing frequency and severity of extreme weather events and the accelerated triggering of climate change tipping points will require a further quantum leap in our understanding of regenerative practices and ways of life. 'Surviving and maybe even turning back the tide of this pervasive ongoing disaster will require more ability to improvise together, stronger societies, more confidence in each other. It will require a world in which we are each other's wealth and have each other's trust' [319, p. 308].

The growing awareness that we are entering a long emergency of constantly rolling, intertwined disasters leads me also to the work of the American science fiction author, NK Jemison [320]. Jemisin's trilogy of *Broken Earth* novels, *The Fifth Season, The Obelisk Gate and The Stone Sky* helps us visualise a world in which human beings must learn the arts and skills required to constantly rebuild communities overwhelmed by frequent, unpredictable and devastating earthquakes and tsunamis; fires and floods; plagues and famines. Sustaining cultures of regeneration and resilience in such harsh and brutal times is made even tougher by awareness of the role which the greed and hubris of previous generations have played in creating this burning, broken Earth.

Alastair Iles' 2019 article, *Repairing the Broken Earth,* identifies four key principles and practices illuminated by Jemison's novels which might help us face the long road in front of us with honesty, hope and courage [321]. In reviewing and reflecting on these principles, I am struck by how closely they align with many of the themes and ideas I have been exploring in this book.

> Recognising and respecting the complexity and fragility of all the peoples, species and ecologies of the Earth.
> Learning the wisdom and humility to overcome our hubris and pay closer attention to the risks and consequences of overstepping the limits of our knowledge and power.
> Creating and sustaining spaces, institutions and systems where we can constantly renew and relearn our understanding of the skills required to flourish in a deeply interwoven interdependent world.
> Redistributing power and wealth in order to reduce the suffering of human beings and other species, repair ecological damage and begin to rebuild systems and relationships of care and justice; sustainability and resilience.

If the *Broken Earth* trilogy stands at the darker end of speculation about ways in which the climate emergency might evolve, Kim Stanley Robinson's engaging and provocative 2020 thought experiment, *Ministry for the Future*, helps us realise that imagining and visualising a just and resilient post-carbon future is not, in fact, our greatest challenge. The even more urgent, even more complex task is designing and building the pathways and bridges enabling us to safely and successfully cross the 'great trench' from here to there [322].

The bridge to the future visualised by Robinson includes a broad array of disruptive technological and societal innovations: homes and factories powered by wind and sun and hydrogen; regenerative, carbon-negative agriculture; rewilding half the Earth to accelerate reforestation and replenish biodiversity; dramatically expanded roles for community and worker co-operatives; heighted awareness of the links between reducing inequality and cutting emissions; carbon taxes and climate risk litigation; direct air capture and draw down of $Co2$; pumping water from under the Arctic ice to slow the sliding of glaciers into the sea.....

Robinson's central, toughest questions focus our attention however on the policies, strategies and alliances required to transform social, political and economic institutions and relationships. What skills and strategies do we need to build and cross the bridge from neoliberal capitalism to a more democratic, just and sustainable zero-carbon world?' [323]. Perhaps even more challenging, what sources of resilience and replenishment can enable each of us continue to contribute to this work while also renewing and sustaining our daily practices of creativity, courage and compassion?

11

The 2050 Zero-Carbon World Oration

Finalising this book in the shadows of the COVID-19 pandemic provides a stark warning against the naivety and hubris of over-confident predictions about the ways in which human societies and ecologies will evolve. At the time of writing, in early 2021 my social media feeds include the full spectrum of utopian and dystopian scenarios. At one end, there are those who see glimpses of renewed respect for the scientifically informed decision-making; socially inclusive public policy and decisive government leadership capable of driving rapid and equitable climate emergency action.

As the dire economic and social consequences of the pandemic become clearer, many others point to disturbing signs that authoritarian totalitarianism and xenophobic nationalism are becoming even more firmly entrenched. In this future, the short-term benefits of temporary emission reductions are quickly overwhelmed with the rich and powerful ramping up investment in the walls and fortresses designed to maximise protection from unabated global warming and increasingly desperate populations. The following backcasting narrative, sitting somewhere in the middle of these views, is offered as a provocation—not a prediction.

Looking back from 2050, this story is told through a fictional Oration delivered by Professor Teuila Apatu, Director of the *Global Institute for Climate and Energy Transitions,* reflecting on the key events, decisions and actions which drove the great twenty-first-century energy transition at remarkable speed and scale. While celebrating and honouring the impressive progress made by 2050 towards achieving the goal of zero net emissions, Professor Apatu also draws our attention to the ongoing ecological damage and human suffering caused by the failure to reduce emissions with sufficient urgency in the first quarter of the twenty-first century. Professor Apatu concludes therefore by calling on her audience to continue to rise to the challenge of further accelerating the drawdown of emissions and undertaking the replenishing and regenerative action required to enable human beings and all other species to continue to thrive and prosper.

Transcript of the 2050 Zero-Carbon World Oration delivered by Professor Teuila Apatu, Director, Global Institute for Climate and Energy Transitions, Auckland, Aotearoa New Zealand

Good evening. I am delighted to be with you tonight to deliver the *2050 Zero-Carbon World Oration.* As you are aware, this annual oration was inaugurated by the UN Secretary General in 2030 in order to celebrate and accelerate progress in creating a just and resilient zero-carbon world.

In preparing tonight's 2050 Oration, I have been acutely aware of the images we are all seeing of the tragic loss of life from the most recent of the climatic catastrophes which continue to sweep across our planet—the Great Inundation of the Ganges Basin. The grim scenes we have seen in the cities and villages of India and Bangladesh over the last few weeks are one more stark reminder of the need to continue to accelerate progress in achieving our three crucial global priorities: creating a just and resilient zero-carbon economy; protecting and assisting the most vulnerable individuals and communities; and repairing and regenerating our climatically disrupted, burned and flooded world.

My task tonight however is to focus on the first of these priorities by addressing two key questions. How, despite facing so many difficult challenges, have we made such remarkable progress towards achieving a zero-carbon global economy? What have been the key turning and tipping points that have driven the global energy transition at such remarkable speed and scale?

I'd like to begin by sharing some brief observations on the way this challenge looked back in 2020, at the beginning of the transition period. Following a brief overview of major energy transition trends and outcomes over the last 30 years, I then outline five key drivers of the Great Energy Transition. My presentation concludes with some reflections on the difficult journey which still lies before us.

The Age of Foolishness or the Age of Wisdom? The View from 2020

By 2020, the key elements of the post-carbon economy roadmap were well understood: rapid reduction in energy demand through improved energy efficiency and reduced consumption; comprehensive electrification of energy supply; rapid replacement of fossil fuels by renewables; low-carbon agriculture and forestry; carefully managed bio-sequestration; and the actions needed to ensure that the energy transition was undertaken in an equitable and resilient way.

There was also widespread recognition that the most significant roadblocks preventing rapid de-carbonisation were social and political rather than technological. These roadblocks included the power and influence of the fossil-fuel industry and other vested interests; political paralysis and denial; social and technological path dependencies; financial, governance and implementation constraints; and the dominant neoliberal economic paradigm of unsustainable consumption and inequitable wealth distribution.

I have four strong memories from those distant times which I'd like to share with you. The first, in Paris in December 2015, was a moment of genuine elation as I joined thousands of delegates in leaping to my feet to applaud the announcement by COP 21 Chair Laurent Fabius that 195 nations had approved the *Paris Climate Agreement*. Most of us at the time recognised that the *Paris Agreement* was far from perfect, sharing journalist George Monbiot's astute analysis that 'by comparison to what it could have been, it's a miracle. By comparison to what it should have been, it's a disaster' [324].

We very well understood that the COP21 national emission reduction commitments would need to be rapidly strengthened if we were to have any hope of keeping global warming below 2 degrees. Many of us were also keenly aware that a strong emphasis on negative emissions would be required to have any real chance of keeping global warming to 1.5 or even 2 degrees. We were however hopeful that ratifying the *Paris Agreement* would send a strong message to political leaders and global investors that the rapidly accelerating global shift from fossil fuels to renewable energy was now unstoppable.

Just twelve months later, in November 2016, at COP22 in Morocco, I vividly recall the shocked faces of delegates receiving the news that Donald Trump had been elected President of the United States. To what extent, we wondered, would the election of this autocratic champion of climate change deniers and fossil-fuel billionaires derail climate and energy action at the very moment at which it needed to be rapidly accelerated? Of even more concern, would the Trump Presidency drive a further deterioration in the erosion of trust in scientific methods and evidence?

My third, most inspirational memory is of the actions and words of the young Swedish school student Greta Thunberg who, on 18 September 2018, sat down on the steps of the Swedish Parliament to begin her strike from school in protest at the lack of climate action by the Swedish Government. Within a few months, Greta's actions had triggered an international *School Strike 4 Climate* movement bringing together millions of school students from around the world demanding that their governments accelerate decisive, emergency speed climate change action. As we now know, many of these students continue to play crucial ongoing roles in the Great Energy Transition.

And then, in early 2020, the coronavirus pandemic began to sweep across the world. The tragic death toll and huge social and economic dislocation are seared into the memories of all who lived through those deeply distressing times. It was, to paraphrase Charles Dickens, a time of risk and a time of possibility; an age of foolishness and an age of wisdom; an epoch of incredulity and an epoch of belief. While in no way wishing to underestimate the scale of suffering experienced by so many individuals and communities, it is now clear that the disruptive impact of COVID-19 did help trigger a series of interlocking social tipping points with profound implications for the direction and speed of the Great Energy Transition.

In the short term, the collapse of global economic growth, the grounding of aircraft and the lockdown of many large cities did lead to a sharp fall in global emissions. While the benefits of these emission reductions were quickly reversed in the rush to restart stalled economies, the pandemic also left a legacy of learning about the desirability and feasibility of car-free streets, localised supply chains, virtual conferencing and reduced air travel. In hindsight, the even more crucial learning was recognition that the societies most successful in dealing with the pandemic were those which took swift and decisive action informed by rigorous scientific advice and underpinned by strong, inclusive public health and community support systems.

Reflecting back on this complex mix of confronting and inspiring recollections also led me to reread some of the writing which I remember finding particularly useful at that time. I'd like to share a few observations from two of these. The *2015 Report, World in Transition: A Global Social Contract for Sustainability* produced by the *German Advisory Council on Global Change* brought together a broad array of research and analysis on the sources and drivers of large-scale technological, social and economic transformations [325]. I remember being particularly struck by the Report's assessment that 'avoidance of dangerous climate change, and the aversion of other threats to humankind as part of the Earth system' would need to go 'far beyond technological and technocratic reforms' and would in fact require the creation of 'a new global social contract for a low-carbon and sustainable global economic system.' This new social contract, the Report argued, would in turn require the creation of a 'culture of attentiveness (born of a sense of ecological responsibility), a culture of participation (as a democratic responsibility), and a culture of obligation towards future generations (future responsibility).'

The second document I'd like to refer you to is *A Roadmap for Rapid Decarbonisation* published in the journal *Science* in March 2017 [54]. This paper, authored by some of the most well-informed climate science analysts of that time, summarised key actions required to create a 40% probability of keeping global warming below 1.5 degree. The article began by stressing the importance of ensuring that emissions peaked by 2020 and then continuing to halve gross CO_2 emissions every decade until 2050. It was, in fact, rising global alarm at the slow progress towards achieving these targets, combined with the tragic impact of several catastrophic extreme weather events that triggered the global citizen movement leading to *Jakarta 2025*.

Milestones on the Journey to a Zero-Carbon World

The handout provided to you (Table 11.1) provides an overview of key energy transition milestones in the last 30 years. It is striking to note that these have generally followed a similar, although slightly slower trajectory to the priority actions proposed in the *Deep Decarbonisation Roadmap* article. Evidence continues to mount that the ongoing implications of this delay have included a faster than hoped for rise in emissions and global temperatures; a higher incidence of extreme weather events; and an increased reliance on negative emission interventions.

Drivers of the Great Energy Transition

While in no way underestimating the enormous climatic challenges still ahead of us, it is, I believe, extremely valuable to remember the key actions and decisions which have driven the Great Energy Transition thus far.

Sustained leadership from national and sub-national governments, business, civil society and the military in ratcheting up and accelerating implementation of the Paris Agreement.

As noted earlier, ratification of the *2015 Paris Agreement* provided a clear, strong signal that governments representing the vast majority of the world's population were firmly committed to decisive climate action. While the commitment of particular national governments has

Table 11.1 Key climate and energy transition milestones 2020–2050

	Extreme weather events and disasters	Key political and policy events and interventions	Energy transition milestones	Emissions trends
2020–2030	2019–2020: COVID-19 Pandemic 2022: Los Angeles Firestorm 2024: Delhi Air Quality Evacuation 2025: Mexico City Water Crisis 2027–2030: China-India 'Water Wars' 2028: Cyclone Kali (Bangla Desh)	2021: Paris Agreement Review further strengthens energy transition goals and targets 2023: Climate Emergency Alliance wins majority in US Congress. US Climate Emergency Act passed 2025: Jakarta Climate and Energy Summit 100 countries commit to 2050 carbon neutral economy Comprehensive decarbonisation plans announced by all major cities and corporations Cap and trade schemes in place in most jurisdictions Carbon price rises to $150 a metric ton Fossil fuel subsidies eliminated Global moratorium on new unabated coal energy	Renewables provide 50% of global energy mix Energy storage and smart grids enable affordable management of renewable energy intermittency issues Leading cities (e.g. Copenhagen, Hamburg, San Francisco) carbon neutral Internal combustion engines in new cars phased out Decarbonisation of long-distance transport through renewable fuels and electrification Rail replaces air traffic for short haul freight and passenger transport Coal exits global energy mix	Land use emissions decrease to 2 $GtCO_2e$ pa Gross CO_2 e emissions decline from 40 to 24 gigatons Bioenergy with Capture and Storage (BECCS) and Direct Air Capture and Storage (DACS) technology removes 0.5 $GtCO_2e$ pa

(continued)

Table 11.1 (continued)

	Extreme weather events and disasters	Key political and policy events and interventions	Energy transition milestones	Emissions trends
2030–2040	2030: Central Africa Drought and food riots 2035: East Coast Australia Firestorm 2038: Collapse of Antarctic Larsen Ice shelf	• Carbon pricing expanded to cover all GHG emissions, including air travel and shipping • Carbon price rises to $300 per metric ton • 2030: Cape Town Climate Justice Summit mobilizes funds to address climate impacts on most vulnerable populations • 2035 Delhi Climate Engineering Summit	• Renewables provide 75% of global energy mix • All building construction carbon neutral (including emissions free steel and concrete) • Electrification of all sectors in lead countries • Phase out of internal combustion engines • Aircraft fuel entirely carbon neutral • Oil exits global energy mix	• Land use emissions 12 $GtCO_2e$ pa • Gross CO_2 emissions decline to 14 gigatons • BECCS and DACS removing 2 $GtCO_2e$ pa
2040–2050	2040: Collapse of Atlantic fishing grounds 2045: Siberian methane explosion 2050: Ganges Basin Inundation	• 2040: Paris Climate Summit celebrates 25 years of progress on achieving Paris Agreement goals. Identifies key ongoing priorities with increased focus on protecting cities and regions most vulnerable to climate impacts • Carbon price: $400 per metric ton • Zero-Carbon World Global Festival, 1 January 2050	• Renewables provide 95% of energy mix • All major European countries and US carbon neutral by 2040 • Electrification of all sectors in most countries • Most other nations carbon neutral by 2050	• Land use emissions decrease to zero • Gross e emissions decline to 5 gigatons • On track for zero CO_2e emissions 2050–2060

varied over time, there has been crucial and sustained leadership from countries responsible for the largest proportion of emissions including China, Germany, India, Indonesia and, after a dreadfully slow start, the United States. International agreement and co-operation to support a rapid rise in the price of carbon (reaching $US400 a tonne in 2040) have been fundamental to the achievement of an energy transition at such an impressive speed and scale.

Sub-national governments and cities, from California to Scotland and from Shanghai to Stockholm, have also played a critical role in demonstrating the feasibility and desirability of zero-carbon pathways and partnerships. The courageous role which the Californian Government played in countering the initial destructive impact of US withdrawal from the Paris Agreement is, of course, well known. Step one was the announcement that California would lead ratification of the Paris Agreement by US and other sub-national jurisdictions. Step two was the formal declaration by the Governor of California of a Climate Emergency and the passage of the Californian Climate Emergency Act, authorising the deployment of public, private and military resources and expertise required to address climate risks and ensure the implementation of emissions reduction actions at the speed and scale fully required to achieve the Paris Agreement targets.

Unprecedented levels of grassroots protest, civil disobedience and demonstrations; the recognition by investors that the economic tipping point from fossil fuel to renewables had now arrived; the impact on the US economy of Chinese, EU and Indian trade sanctions were clearly all important, as was climate and energy leadership from high-ranking business, military, public sector and community leaders. Most important of all I continue to believe was the leadership provided by wave after wave of students and young people carrying on the work of the initial leaders of *School Strike 4 Climate*.

The mobilisation of millions of citizens imagining and creating a diverse array of fossil-fuel divestment, consumer boycott, trade sanction and civil disobedience resistance strategies.

In hindsight, the massive demonstrations and protest movements such as the *People's March for Climate* and the *Sunrise Movement* which erupted across the United States in response to the Trump administration's assault on climate action were just the first wave of a far broader global mobilisation. The 2020 election of President Joe Biden and the decision by the United States to rejon the Paris Agreement were important turning points enabling some renewal of global climate action momentum. The truly transformational moment in US politics finally arrived in 2023 with the Congressional vote to approve financing for the *Climate Emergency Act* and the emergency speed implementation of the *Green New Deal*, the comprehensive package of measures designed to trigger an emergency speed decarbonisation of the US economy. As the eminent climate justice and human rights advocate Alexandria Ortasio-Cortez correctly noted with considerable prescience, 'this is going to be the Great Society, the moonshot, the civil rights movement of our generation' [102].

The Escalating Frequency and Severity of Catastrophic Climatic Events

Tragically, the history of the first half of the twenty-first century is as much a history of catastrophic climate-driven disasters as it is of transformational change in social, economic and technological systems. The devastating impact of cyclones Katrina, Tracy and Haiyan; the great droughts in Somalia and California; and the heatwaves and wildfires which swept Australia, Chile, Canada and Russia were only the first of many climate wake up calls which too many of us continued to ignore for too long.

Finally, however, the escalating frequency and severity of extreme weather events overwhelmed the carefully constructed defences of our wishful thinking and denial, not to mention the patience of investors in the world's largest insurance companies. The terrible bushfires that swept the East Coast of Australian in 2019; the Los Angeles Firestorm (2022); the Delhi air quality evacuation (2024); the Mexico City water crisis (2025); the India-China 'Water Wars' (2027–2030); the Central African 'Famine Wars' (which reached their darkest days in 2030); and the collapse of the Atlantic fishing grounds all played key roles in strengthening support for further acceleration of the energy transition. Increasing understanding of the link between extreme weather events, food insecurity, refugee flows and military conflict also further strengthened support for the adaptation and resilience investments needed to protect vulnerable populations.

Disruptive, Game-Changing Technological Innovation

The most visible drivers of transformational change have often been technological: cascading and disruptive innovations in energy efficiency; solar, wind, tidal and geothermal energy; energy storage (batteries and pumped hydro); electrified and autonomous transport systems; startling breakthroughs in aviation biofuels, low-carbon construction materials and digital fabrication have all played important roles as have new technologies for recording and exchanging value such as bitcoin and blockchain.

Smart grids and integrated transmission networks have significantly enhanced the scale and efficiency of energy distribution systems. Heroically ambitious engineering projects linking renewable energy producers with consumers in North Africa and Europe; North and South America; and Australia and South-East Asia have all made major contributions as have impressive improvements in long-distance high-voltage DC electricity transmission. At more local scales, urban smart grids have enabled and accelerated lateral energy trading between local households and businesses.

Dramatic improvements in soil science, livestock feeding practices, forestry and savannah management have all contributed to reducing land use emissions. These advances have been augmented by a slow but continuing cultural shift towards reduced demand for meat, a decrease in the wastage of food and an increase in consumption of locally grown food.

And then there is the profoundly important—and profoundly troubling question—of negative emissions. Thirty years ago, I led debates examining the ecological and political risks of overreliance on Bioenergy with Carbon Capture and Storage (BECCS) technologies. My key concern, which I still believe was correct, was that, even if it was possible to deliver large-scale BECCS solutions at affordable cost, the implications for the capacity of the planet to produce sufficient biomass for human and animal consumption would be unacceptable. BECCS solutions were also consistently being inserted into emissions reduction scenarios as a 'get out of jail card' to avoid tough decisions about changes in lifestyle and consumption.

Both these concerns have turned out to be well founded. It is now very clear that the failure to act with sufficient speed to reduce emissions in the first half of this century has had the long-term effect of further intensifying arguments for expanding investment in BECCS as well as further intensifying pressure on the biodiversity and carrying capacity of the Earth's eco-systems. I can also not finish this section of my presentation without mentioning the most bitter of all Climate Summit debates, the *2035 Delhi Climate Engineering Summit* which led to the very belated, extremely grudging approval for transparent monitoring, evaluation and governance of large-scale geoengineering initiatives such as atmospheric aerosol injections and ocean fertilisation.

Disruptive Game-Changing Innovation in Social, Economic and Political Systems

I have spent my whole life working shoulder to shoulder with the most brilliant engineers and scientists of my generation. I have the highest admiration for the ingenuity and dedication they have brought to the mission of designing and building the extraordinary, game-changing technologies we have seen brought to scale over the last fifty years. My personal view remains however that the most influential drivers of the great energy transition have been radical and disruptive transformations in social and economic systems as well in the cultural values and political institutions. Let me highlight five of the key trends in social, economic and political values and practices which have, in my view, been particularly significant.

Increasing recognition that reducing the global consumption of goods and services is an essential basis for decreasing energy demand and addressing sustainability and climate change challenges. There has also been a belated and contested, but, in my view, crucial shift in understanding that the long working hours and high stress lifestyles required to accumulate an endlessly expanding wish list of material possessions are incompatible with sustained improvements in health and wellbeing. This realisation has triggered and been informed by multiple experiments exploring localised, low consumption systems of energy, food, transport and housing production and consumption. In thinking about the distance, we still have to travel on the road to a more sustainable economic paradigm I am often drawn to this reflection from an old Bhutanese friend of mind. 'When will enough us of learn that expanding time with our friends and family and for explore our creative life is far more rewarding than more and more time at work further expanding the size of our houses and the speed of our cars?'

Increasing popular support for the global climate justice movement, informed by growing awareness that ever-increasing levels of inequality are economically counterproductive, socially corrosive and ecologically unsustainable as well as ethically unjust. These ideas and arguments first canvassed in the *2002 Bali Climate Justice Principles* finally achieved widespread support at the *2030 Cape Town Climate Justice Summit*, which approved comprehensive new measures mobilising funds to address the impact on the most vulnerable populations and communities of climate and energy transitions.

An ongoing shift towards more distributed and collaborative economic paradigms and systems characterised and driven by open-source, peer-to-peer networks of technological knowledge and skills. As a number of thoughtful analysts have usefully noted the 'age of information and telecommunications' which began to emerge in the early twenty-first century can be usefully understood as a 'fifth revolution' in techno-economic paradigms, building on four earlier 'ages': the first industrial revolution (characterised by the introduction of mechanised production systems); the age of steam and railways; the age of steel, electricity and heavy engineering; and the age of oil, automobiles and mass production. This transition to a more collaborative and distributed economy has created enormous opportunities for the acceleration of the knowledge and innovation driving the Great Energy Transition.

Radical innovation in governance arrangements improving the transparency and accountability of economic and political institutions and relationships. Legal and regulatory interventions mandating far tougher corporate standards for the transparent disclosure of climate change and energy transition risks were a crucial driver in shifting investment from fossil fuel to renewable energy industries. At the same time, increasing concern about the political influence of vested interests (particularly in the fossil-fuel industry) and a corrosive decline in public trust for politicians led many governments to significantly tighten and, in many cases, ban outright corporate donations to political parties.

A Great Leap Forward on a Long and Challenging Road

In concluding this speech, I am delighted to announce that the *2050 World Energy Report* to be released tomorrow will show that we are indeed on track to achieve a net zero-carbon economy by 2060. This is an extraordinary achievement and one which I must say has far exceeded my most optimistic expectations in the difficult years following the signing of the *Paris Agreement*.

Many profound challenges and tough questions however remain. Despite the remarkable speed of the Great Energy Transition, further acceleration in negative emissions innovation will clearly be required to bring long-term global warming trends back below 1.5 degrees. The question we still have no answer to is: How can negative emissions at this scale be achieved without overwhelming the capacity of the biosphere to feed nine billion people?

Despite the progress towards the achievement of a zero-carbon global economy, too many crucial economic decisions with profound implications for the future lives of all the species on the planet are still made in ways which are largely invisible and unaccountable. The complex social and cultural pathways leading to a genuinely sustainable, genuinely equitable post-growth economy have yet to be traversed.

Despite the passionate commitment and creativity of so many inspiring scientific, political, business and community leaders over the last 50 years, I am still unable to provide my grandchildren the simple, essential gift I most wish to give them: the ecological conditions which enable human beings to continue to thrive and prosper—alongside the many species with which we share this extraordinary planet.

The rapidity of the twenty-first-century energy transition does provide a powerful illustration of the potential for human imagination and ingenuity combined with ethical and visionary leadership to drive transformational change at remarkable scale and speed. The task which now lies before us is to continue to explore and address the questions raised in a book called *Climate Crisis, Hope and Courage*, which I remember first reading way back in the dark, bleak winter of 2021. What sources of wisdom and insight can strengthen our capacity to take courageous and effective action and to live meaningful and creative lives in a world of rapidly accelerating climatic and ecological risks?

Wherever you are the world I look forward continuing our conversation about these vital questions and in working with you all on the next steps in meeting the great challenges which lie before us.

Thank you. I wish you and all your friends and family good fortune and good health in the months and years to come.

12

The Laughter of Children, the Roar of the Ocean: Concluding Thoughts and Questions

The questions and ideas explored in this book have been informed by many hundreds of conversations with family, friends and colleagues about hope and courage; meaning and purpose, wisdom and action in an increasingly harsh and threatening climate. Some of my earliest conversations, back in 2005, were with climate scientists. Here, they said, have a close look at these graphs. Look at how fast these emissions are rising and how fast the ice is melting. And if you find our data a little dry, we suggest you watch this hurricane forming in the Gulf of Mexico. Katrina it's called. Hurricane Katrina.

The images of families stranded on the rooves of their flooded homes as rising water surged through New Orleans were certainly confronting. As were the thousands of deaths from the unprecedented European heat waves of 2006 and 2007. The 2007 IPCC 4th Assessment report confirming the role of human activity in triggering catastrophic climate tipping points. Cyclone Nargis sweeping through Bangladesh and Myanmar in 2008, killing over 100,000 people. Then in February 2009, far closer to home the searing, 45 degree heat of Black Saturday with bushfires roaring through the hills and towns just north of Melbourne.

OK, I agreed, climate action was clearly an urgent priority and an essential basis for handing on to our grandchildren a world at least as full of opportunities as the planet full of wonders on which I have been fortunate enough to be born. If we did not take decisive action by 2015—perhaps 2020 at the latest—we would surely be in deep trouble.

My questions at that time were focused most of all on the possibility of swift and decisive action. Did we have the technological and financial capacity to reduce emissions at the scale and speed required to prevent catastrophic climate change? The answer to this question was clearly yes. Renewable energy from the sun and wind replacing coal and gas and oil. Energy efficiency and electrification driving down energy demand and emissions from our cars and industries and houses. Regenerative farming and low carbon land use. The news all seemed extremely positive with cost of all these possibilities falling at remarkable speed. Except of course that emissions kept rising, global temperatures kept increasing and the storms and floods and fires kept getting worse.

The focus of my conversations began to shift from technological and financial to social and political challenges and obstacles. How could we avoid being overwhelmed and paralysed as we came to fully understand the wealth and power of the mining, media and finance corporations fuelling and driving the politics of climate action denial and delay? And how could we align strategies for overcoming injustice and oppression with the actions and timetables required to achieve emergency speed emission reductions? The responses from climate activists, scientists and policy makers were again remarkably consistent. Visionary and courageous leadership. Skilful communication of the most rigorous scientific evidence. The election of governments firmly committed to swift and decisive action. Massive, broadly based citizen mobilisation and civil disobedience. Disruptive divestment and transformational technological and social innovation.

All of these ideas and strategies will surely continue to make crucial contributions to further accelerating the transition to a just and resilient zero-carbon economy. None of them have however yet triggered transformational change at sufficient speed and scale. Global temperature increases of 4 degrees and more are rapidly coming into view. So now that we have arrived in this age of consequences, we face another daunting question. What sources of wisdom and insight can strengthen our capacity to take courageous and effective action and to live meaningful and creative lives in a world of rapidly accelerating climatic and ecological risks?

There are times, in reflecting on my responses to this question and in visualising the content of this book when I imagine all the friends and colleagues; scientists and activists; teachers and writers; poets and artists whose ideas and voices I have drawn on gathered together in respectful and intense debate. All the speakers are passionate and well-informed; the conversations spark and crackle with fierce, urgent energy.

I turn first to my friends and colleagues from Indigenous and First Nation communities. We might usefully begin, they note by remembering and honouring the histories of the lands on which we gather; the stories of our people and the legacies of colonialism and dispossession which have led us to this place. Climate justice is therefore one of the first propositions we should bring to the table. For the principle of climate justice to be more than hollow words, we will need to see substantive actions which fully acknowledge and address the sources and consequences of violence and injustice. Principles and practices of care, compassion and respect will also be foundational: care and respect for country, for all the creatures with whom we share this world and for all the human beings who will follow after us.

While acknowledging the wisdom of this opening contribution, my colleagues from the world of science and technology approach the question from another direction. We would like, they say to foreground the power of scientific evidence analysed through rigorous research and knowledge crystallised from data in the crucible of critical reason. Speaking truth to power with honesty and skill about the causes and consequences of the threats we face and the actions we need to take to overcome them. All of this sounds eminently reasonable. But how I wonder do they maintain their own emotional resilience in facing tough truths about the future which their evidence tells them is increasingly likely? Here the responses are more complicated and more varied. Remembering and trusting in the disruptive, game-changing power of ingenuity, creativity and innovation are sometimes enough. And also dancing, meditation, art and music; working together in the garden; the kindness, warmth and joy of family and friends; walking the old dog by the ocean, watching the horizon as the waves keep rolling in.

My climate activist friends seem less convinced by the promise and power of reason and innovation. Listen to the science and accelerate technological innovation. Of course. There are indeed many impressive examples of the human capacity for creativity and inventiveness to overcome disease and hunger, suffering and injustice. But how do we deploy data and evidence and reason with the speed and skill required to accelerate just and inclusive emission reduction strategies while avoiding the delusional hubris that there are always technical solutions to every problem? The historical examples which my activist colleagues turn to most of all for encouragement and inspiration are stories of solidarity and fellowship; comradeship and reciprocity where ethically informed collective action has achieved transformational change which once looked completely impossible. The anti-slavery movement, the *Suffragettes,* the overthrow of Apartheid, the fall of the Berlin Wall. More recently *School Strike 4 Climate, Pacific Climate Warriors, Extinction Rebellion, The Sunrise Movement and Black Lives Matter.* And also, crucially the struggles of Indigenous peoples in Australia and New Zealand, Canada and the United States, Bolivia and Brazil for land rights, justice and self-determination.

I am joined then by teachers and scholars from a wide array of spiritual and faith-based traditions and perspectives. The first, foundational steps they suggest in times of suffering and despair are thankfulness and gratitude. Honouring and celebrating the astonishing, complex beauty of life on Earth is an abiding source of strength and inspiration. Awareness and understanding of the fragile impermanence of our dew drop world is also a constant reminder of our shared responsibility to keep paying attention; to keep turning up; to hold the line and to keep nurturing and sustaining relationships and practices of kindness and compassion; justice, love and care.

I turn finally to the critical theorists and writers; artists and designers who can help us imagine and create the ecologically informed paradigms and practices of resilience and regeneration we will need to navigate the wild landscapes of the long emergency. Good companions who can help us to more clearly see the patterns and textures of our interwoven world and to understand and confront the ignorance, violence and greed threatening to tear this delicate fabric apart. Experienced guides who can assist us make well-informed and well-considered choices about pathways we should choose and places we should land. Thoughtful teachers who can help us learn the art of living well in dangerous, uncertain times, remembering that the world is always full of surprises and the future is never entirely settled.

These then are some of the ideas and insights; strategies and practices which help me answer the questions I have set myself. What sources of wisdom can strengthen our capacity to take courageous and effective action and to live meaningful lives in a world of rapidly accelerating climatic and ecological risks? Emergency speed, science-based collective action. Justice and care; respect and reciprocity. Reason, ingenuity and technology. Attentiveness and thankfulness, kindness and compassion. Joyfulness and beauty; creativity and imagination. And also these abiding gifts: the laughter of children; the comfort of old friends; sunlight on the water; the wind in the trees; the silence of mountains; the roar of the ocean.

Appendix A: Donella Meadows, Leverage Points—Places to Intervene in a System

This summary by Donnella Meadows of leverage points with the potential to bring about systemic change is deliberately presented by her in reverse order. The potential impact and power of the interventions increases as the reader reads down through the list. *Source* Donella Meadows. (2008). *Thinking in Systems: A Primer.* Chelsea Green Publishing (pp. 145–165).

12. Numbers: Constants and parameters such as subsidies, taxes, standards
11. Buffers: The sizes of stabilising stocks, relative to their flows
10. Stock and flow structures: Physical systems and their nodes of intersection
9. Delays: The lengths of time relative to the rates of system changes
8. Balancing feedback loops: The strength of feedbacks relative to the impacts they are trying to correct
7. Reinforcing feedback loops: The strength of the gain of driving loops
6. Information flows: The structure of who does and does not have access to information
5. Rules: Incentives, punishments and constraints

4. Self organisation: The power to add, change or evolve system structure
3. Goals: The purpose or function of the system
2. Paradigms: The mindset out of which the system including its goals, structure, rules, delays, parameters arises
1. Transcending paradigms

Appendix B: Bali Principles of Climate Justice, 2002

Source Bali Principles of Climate Justice. (2002, August 28). http://www.indiaresource.org/issues/energycc/2003/baliprinciples.html.

1. Affirming the sacredness of Mother Earth, ecological unity and the interdependence of all species, Climate Justice insists that communities have the right to be free from climate change, its related impacts and other forms of ecological destruction.
2. Climate Justice affirms the need to reduce with an aim to eliminate the production of greenhouse gases and associated local pollutants.
3. Climate Justice affirms the rights of indigenous peoples and affected communities to represent and speak for themselves.
4. Climate Justice affirms that governments are responsible for addressing climate change in a manner that is both democratically accountable to their people and in accordance with the principle of common but differentiated responsibilities.
5. Climate Justice demands that communities, particularly affected communities, play a leading role in national and international processes to address climate change.

6. Climate Justice opposes the role of transnational corporations in shaping unsustainable production and consumption patterns and lifestyles, as well as their role in unduly influencing national and international decision making.
7. Climate Justice calls for the recognition of a principle of ecological debt that industrialised governments and transnational corporations owe the rest of the world as a result of their appropriation of the planet's capacity to absorb greenhouse gases.
8. Affirming the principle of ecological debt, Climate Justice demands that fossil fuel and extractive industries be held strictly liable for all past and current life-cycle impacts relating to the production of greenhouse gases and associated local pollutants.
9. Affirming the principle of Ecological debt, Climate Justice protects the rights of victims of climate change and associated injustices to receive full compensation, restoration and reparation for loss of land, livelihood and other damages.
10. Climate Justice calls for a moratorium on all new fossil fuel exploration and exploitation; a moratorium on the construction of new nuclear power plants; the phase out of the use of nuclear power world wide; and a moratorium on the construction of large hydro schemes.
11. Climate Justice calls for clean, renewable, locally controlled and low-impact energy resources in the interest of a sustainable planet for all living things.
12. Climate Justice affirms the right of all people, including the poor, women, rural and indigenous peoples, to have access to affordable and sustainable energy.
13. Climate Justice affirms that any market-based or technological solution to climate change, such as carbon-trading and carbon sequestration, should be subject to principles of democratic accountability, ecological sustainability and social justice.
14. Climate Justice affirms the right of all workers employed in extractive, fossil fuel and other greenhouse gas producing industries to a safe and healthy work environment without being forced to choose between an unsafe livelihood based on unsustainable production and unemployment.

15. Climate Justice affirms the need for solutions to climate change that do not externalise costs to the environment and communities, and are in line with the principles of a just transition.
16. Climate Justice is committed to prevent the extinction of cultures and biodiversity due to climate change and its associated impacts.
17. Climate Justice affirms the need for socio-economic models that safeguard the fundamental rights to clean air, land, water, food and healthy ecosystems.
18. Climate Justice affirms the rights of communities dependent on natural resources for their livelihood and cultures to own and manage the same in a sustainable manner, and is opposed to the commodification of nature and its resources.
19. Climate Justice demands that public policy be based on mutual respect and justice for all peoples, free from any form of discrimination or bias.
20. Climate Justice recognises the right to self-determination of Indigenous peoples, and their right to control their lands, including subsurface land, territories and resources and the right to the protection against any action or conduct that may result in the destruction or degradation of their territories and cultural way of life.
21. Climate Justice affirms the right of indigenous peoples and local communities to participate effectively at every level of decision making, including needs assessment, planning, implementation, enforcement and evaluation, the strict enforcement of principles of prior informed consent, and the right to say 'No.'
22. Climate Justice affirms the need for solutions that address women's rights.
23. Climate Justice affirms the right of youth as equal partners in the movement to address climate change and its associated impacts.
24. Climate Justice opposes military action, occupation, repression and exploitation of lands, water, oceans, peoples and cultures, and other life forms, especially as it relates to the fossil fuel industry's role in this respect.
25. Climate Justice calls for the education of present and future generations, emphasises climate, energy, social and environmental issues,

while basing itself on real-life experiences and an appreciation of diverse cultural perspectives.

26. Climate Justice requires that we, as individuals and communities, make personal and consumer choices to consume as little of Mother Earth's resources; conserve our need for energy; and make the conscious decision to challenge and reprioritise our lifestyles, rethinking our ethics with relation to the environment and the Mother Earth, while utilising clean, renewable, low impact energy; and ensuring the health of the natural world for present and future generations.

27. Climate Justice affirms the rights of unborn generations to natural resources, a stable climate and a healthy planet.

Appendix C: Universal Declaration of Rights of Mother Earth, World People's Conference on Climate Change and the Rights of Mother Earth, Cochabamba, Bolivia April 2010

Source World People's Conference on Climate Change and the Rights of Mother Earth Peoples Agreement. (2010, January 5). https://pwccc.wordpress.com/support/.

Preamble

We, the peoples and nations of Earth:

considering that we are all part of Mother Earth, an indivisible, living community of interrelated and interdependent beings with a common destiny;

gratefully acknowledging that Mother Earth is the source of life, nourishment and learning and provides everything we need to live well;

recognising that the capitalist system and all forms of depredation, exploitation, abuse and contamination have caused great destruction, degradation and disruption of Mother Earth, putting life as we know it today at risk through phenomena such as climate change;

convinced that in an interdependent living community it is not possible to recognise the rights of only human beings without causing an imbalance within Mother Earth;

affirming that to guarantee human rights it is necessary to recognise and defend the rights of Mother Earth and all beings in her and that there are existing cultures, practices and laws that do so;

conscious of the urgency of taking decisive, collective action to transform structures and systems that cause climate change and other threats to Mother Earth;

proclaim this Universal Declaration of the Rights of Mother Earth, and call on the General Assembly of the United Nation to adopt it, as a common standard of achievement for all peoples and all nations of the world, and to the end that every individual and institution takes responsibility for promoting through teaching, education, and consciousness raising, respect for the rights recognised in this Declaration and ensure through prompt and progressive measures and mechanisms, national and international, their universal and effective recognition and observance among all peoples and States in the world.

Article 1. Mother Earth

(1) Mother Earth is a living being.
(2) Mother Earth is a unique, indivisible, self-regulating community of interrelated beings that sustains, contains and reproduces all beings.
(3) Each being is defined by its relationships as an integral part of Mother Earth.
(4) The inherent rights of Mother Earth are inalienable in that they arise from the same source as existence.
(5) Mother Earth and all beings are entitled to all the inherent rights recognised in this Declaration without distinction of any kind, such as may be made between organic and inorganic beings, species, origin, use to human beings or any other status.
(6) Just as human beings have human rights, all other beings also have rights which are specific to their species or kind and appropriate for

their role and function within the communities within which they exist.
(7) The rights of each being are limited by the rights of other beings and any conflict between their rights must be resolved in a way that maintains the integrity, balance and health of Mother Earth.

Article 2. Inherent Rights of Mother Earth

(1) Mother Earth and all beings of which she is composed have the following inherent rights:
   (a) the right to life and to exist;
   (b) the right to be respected;
   (c) the right to regenerate its bio-capacity and to continue its vital cycles and processes free from human disruptions;
   (d) the right to maintain its identity and integrity as a distinct, self-regulating and interrelated being;
   (e) the right to water as a source of life;
   (f) the right to clean air;
   (g) the right to integral health;
   (h) the right to be free from contamination, pollution and toxic or radioactive waste;
   (i) the right to not have its genetic structure modified or disrupted in a manner that threatens its integrity or vital and healthy functioning;
   (j) the right to full and prompt restoration for violations of the rights recognised in this Declaration caused by human activities;
(2) Each being has the right to a place and to play its role in Mother Earth for her harmonious functioning.
(3) Every being has the right to wellbeing and to live free from torture or cruel treatment by human beings.

Article 3. Obligations of human beings to Mother Earth

(1) Every human being is responsible for respecting and living in harmony with Mother Earth.
(2) Human beings, all States, and all public and private institutions must:

   (a) act in accordance with the rights and obligations recognised in this Declaration;
   (b) recognise and promote the full implementation and enforcement of the rights and obligations recognised in this Declaration;
   (c) promote and participate in learning, analysis, interpretation and communication about how to live in harmony with Mother Earth in accordance with this Declaration;
   (d) ensure that the pursuit of human wellbeing contributes to the wellbeing of Mother Earth, now and in the future;
   (e) establish and apply effective norms and laws for the defence, protection and conservation of the rights of Mother Earth;
   (f) respect, protect, conserve and where necessary, restore the integrity, of the vital ecological cycles, processes and balances of Mother Earth;
   (g) guarantee that the damages caused by human violations of the inherent rights recognised in this Declaration are rectified and that those responsible are held accountable for restoring the integrity and health of Mother Earth;
   (h) empower human beings and institutions to defend the rights of Mother Earth and of all beings;
   (i) establish precautionary and restrictive measures to prevent human activities from causing species extinction, the destruction of ecosystems or the disruption of ecological cycles;
   (j) guarantee peace and eliminate nuclear, chemical and biological weapons;

(k) promote and support practices of respect for Mother Earth and all beings, in accordance with their own cultures, traditions and customs;
(l) promote economic systems that are in harmony with Mother Earth and in accordance with the rights recognised in this Declaration.

Article 4. Definitions

(1) The term 'being' includes ecosystems, natural communities, species and all other natural entities which exist as part of Mother Earth.
(2) Nothing in this Declaration restricts the recognition of other inherent rights of all beings or specified beings.

References

1. Thunberg, G. (2019, September 23). Speech to U.N. climate action summit. *NPR.org.* Retrieved 1 June 2020, from https://www.npr.org/2019/09/23/763452863/transcript-greta-thunbergs-speech-at-the-u-n-climate-action-summit.
2. Wiseman, J., Edwards, T., & Luckins, K. (2013). Post carbon pathways: A meta-analysis of 18 large-scale post carbon economy transition strategies. *Environmental Innovation and Societal Transitions, 8*, 76–93. https://doi.org/10.1016/j.eist.2013.04.001.
3. Frantz, C. M., & Mayer, F. S. (2009). The emergency of climate change: Why are we failing to take action? *Analyses of Social Issues and Public Policy, 9*(1), 205–222. https://doi.org/10.1111/j.1530-2415.2009.01180.x.
4. IPCC. (2014). Climate change 2013: The physical science basis. *Headline Statements from the Summary for Policy Makers Wkg Gp 1 Technical Support Unit.* Cambridge, UK and New York, NY, USA: IPCC.
5. Australian Psychological Society. (2019). *Coping with climate change distress.* Retrieved 7 September 2020 from https://www.psychology.org.au/About-Us/What-we-do/advocacy/Advocacy-social-issues/Enviro

nment-climate-change-psychology/Resources-for-Psychologists-and-others-advocating/Coping-and-adapting-to-climate-change.
6. Williams, R. (1989). *Resources of Hope: Culture, democracy, socialism.* London and New York: Verso.
7. Nietzsche, F. W., & Harvey, A. (2013). *Human, all too human: A book for free spirits.* Auckland, New Zealand: The Floating Press.
8. Stoknes, P. E. (2015). *What we think about when we try not to think about global warming: toward a new psychology of climate action.* White River Junction, Vermont: Chelsea Green Publishing.
9. Jensen, D. (2015, July 1). Give up hope: It's the best chance we have to save everything we love. *Films For Action.* Retrieved 18 December 2017, from http://www.filmsforaction.org/articles/give-up-hope-its-the-best-chance-we-have-to-save-everything-we-love/.
10. Solnit, R. (2016). *Hope in the dark: Untold histories, wild possibilities.* Chicago: Haymarket Books.
11. Solnit, R. (2016, July 15). 'Hope is an embrace of the unknown': Rebecca Solnit on living in dark times. *The Guardian.* Retrieved 1 July 2019 from https://www.theguardian.com/books/2016/jul/15/rebecca-solnit-hope-in-the-dark-new-essay-embrace-unknown.
12. Mazur, Laurie. (2019, July 22). Despairing about the climate crisis? Read this. *Earth Island Journal.* Retrieved 6 July 2020, from https://www.earthisland.org/journal/index.php/articles/entry/despairing-about-climate-crisis/.
13. Macy, J., & Johnstone, C. (2012). *Active Hope: How to face the mess we're in without going crazy.* Novato, CA: New World Library.
14. Lear, J. (2006). *Radical hope: Ethics in the face of cultural devastation.* Cambridge, MA: Harvard University Press.
15. Oxford English Dictionary. (2018). Retrieved 15 November 2018, from http://www.oed.com/.
16. Mandela Nelson. (1989). Mandela, N. Speech to US Congress, 1989. Retrieved 11 June 2018 from https://www.youtube.com/watch?v=Z54jYc1hJfE.
17. May, R. (1975). *The Courage to Create.* New York: Norton [1975].
18. McGregor, J. (2014, May 18). Maya Angelou on leadership, courage and the creative process. *The Washington Post.* Retrieved 16 December 2018, from https://www.washingtonpost.com/news/on-leadership/wp/2014/05/28/maya-angelou-on-leadership-courage-and-the-creative-process/?noredirect=on&utm_term=.16a0a5644ccb.

19. Marvel, K. (2018). 'We need courage, not hope to face climate change, *On Being* https://onbeing.org/blog/kate-marvel-we-need-courage-not-hope-to-face-climate-change.
20. Pope Francis. (2015). *Encyclical on climate change & inequality: On care for our common home.* Brooklyn: Melville House.
21. Fiedorczuk, J. (2014). Can poetry save the earth? An interview with Gary Snyder. *Polish Journal of American Studies, 8,* 7–17.
22. Duggan, J. (2018). Is this how you feel? *Is this how you feel?* Retrieved 26 November 2018, from http://www.isthishowyoufeel.com/.
23. Baldwin, J. (1962, January 14). As much truth as one can bear. *The New York Times Book Review.*
24. Jasanoff, S. & Hilton,R.S. (2017). No Funeral Bells: Public Reason in a 'Post-Truth' Age. *Social Studies of Science, 47.5 (October 2017),* 751-770. https://doi.org/10.1177/0306312717731936.
25. World Meteorological Organization. (2019). United In Science: *High-level synthesis report of latest climate science information convened by the Science Advisory Group of the UN Climate Action Summit 2019* (p. 28). Retrieved from public.wmo.int/en/resources/united_in_science.
26. World Meteorological Organization. (2019, September 19). United in Science 2020. *World Meteorological Organization.* Retrieved 10 November 2020, from https://public.wmo.int/en/resources/united_in_science.
27. Sano, Y. (2013, November 11). "It's time to stop this madness" - Philippines plea at UN climate talks. *Climate Home News.* Retrieved 20 November 2018, from http://www.climatechangenews.com/2013/11/11/its-time-to-stop-this-madness-philippines-plea-at-un-climate-talks/.
28. Jetnil-Kijiner, K. (2017). *Iep Jaltok: Poems from a Marshallese Daughter.* Tucson: University of Arizona Press.
29. Flanagan, R. (2019, February 4). Tasmania is burning. *The Guardian.* Retrieved 2 May 2019 from https://www.theguardian.com/environment/2019/feb/05/tasmania-is-burning-the-climate-disaster-future-has-arrived-while-those-in-power-laugh-at-us.
30. Carrington, D. (2018, December 3). David Attenborough: Collapse of civilisation is on the horizon. *The Guardian.* Retrieved 6 May 2020 from https://www.theguardian.com/environment/2018/dec/03/david-attenborough-collapse-civilisation-on-horizon-un-climate-summit.
31. United Nations Framework Convention on Climate Change (2015) The Paris Agreement. *UNFCCC.* Retrieved 26 November 2018, from https://unfccc.int/process/the-paris-agreement/what-is-the-paris-agreement.

32. Climate Action Tracker. (2018). *Global temperatures*. Retrieved 29 January 2018, from http://climateactiontracker.org/global.html.
33. Myhre, G., & Samset, B. H. (2015). Standard climate models radiation codes underestimate black carbon radiative forcing. *Atmospheric Chemistry and Physics, 15*(5), 2883–2888. https://doi.org/10.5194/acp-15-2883-2015.
34. O'Neill, B. C., Oppenheimer, M., Warren, R., Hallegatte, S., Kopp, R. E., Pörtner, H. O., Yohe, G. (2017). IPCC reasons for concern regarding climate change risks. *Nature Climate Change, 7*(1), 28–37. https://doi.org/10.1038/nclimate3179.
35. Christoff, P. (2011, August 5). Are you ready for a four degree world? *The Conversation*. Retrieved 26 November 2018, from http://theconversation.com/are-you-ready-for-a-four-degree-world-2452.
36. Kanter, J. (2009, March 13). Scientist: Warming could cut population to 1 billion. *Dot Earth Blog*. Retrieved 8 September 2020 from https://dotearth.blogs.nytimes.com/2009/03/13/scientist-warming-could-cut-population-to-1-billion/.
37. Zieler, C. (2009, March 5). Russian roulette odds … If we're lucky. *Uniavisen*. Retrieved 26 November 2018, from https://uniavisen.dk/en/russian-roulette-odds-if-were-lucky/.
38. Anderson, K. (2011, July). *Professor Kevin Anderson - Climate change: Going beyond dangerous*. News & Politics. Retrieved 1 October 2019 from https://www.slideshare.net/DFID/professor-kevin-anderson-climate-change-going-beyond-dangerous.
39. Warren, R. (2011). The role of interactions in a world implementing adaptation and mitigation solutions to climate change. *Philosophical Transactions of the Royal Society A: Mathematical, Physical and Engineering Sciences, 369*(1934), 217–241. https://doi.org/10.1098/rsta.2010.0271.
40. Dlouhy, J. and Wingrove, J. (2020, December 20). Biden introduces his environment team, calling climate change "the existential threat of our time." *Fortune*. Retrieved 22 December 2020, from https://fortune.com/2020/12/19/biden-environment-climate-change-nominees-epa-interior-haaland-michael-regan/.
41. Bostrom, N. (2002). Existential risks: Analysing human extinction scenarios and related hazards. *Journal of Evolution and Technology* (9). Retrieved 8 September 2020 from https://jetpress.org/volume9/risks.html.
42. Bostrom, N., & Cirkovic, M. (2008). *Global Catastrophic Risks*. Oxford, UK: Oxford University Press.

43. Hulme, M. (2009). *Why we disagree about climate change: Understanding controversy, inaction and opportunity.* Cambridge; New York: Cambridge University Press.
44. Gardiner, S. M. (2011). *A perfect moral storm: The ethical tragedy of climate change.* New York: Oxford University Press.
45. Lamb, W. F., Mattioli, G., Levi, S., Roberts, J. T., Capstick, S., Creutzig, F., Steinberger, J. K. (2020). Discourses of climate delay. *Global Sustainability, 3,* e17. https://doi.org/10.1017/sus.2020.13.
46. Oreskes, N., & Conway, E. M. (2010). *Merchants of Doubt: How a handful of scientists obscured the truth on issues from tobacco smoke to global warming.* New York: Bloomsbury Press.
47. Mathiesen, K. (2017, October 10). Tony Abbott says climate change is "probably doing good". *The Guardian Australia.* Retrieved 26 November 2018 from https://www.theguardian.com/australia-news/2017/oct/10/tony-abbott-says-climate-change-is-probably-doing-good.
48. Steffen, A. (2017, March 8). Climate gradualism—"We all want to act on climate, but we also have to be slow, incremental and realistic"—Is the new climate denialism. *@alexsteffen.* Tweet. Retrieved 14 January 2019 from https://twitter.com/alexsteffen/status/839557070399512576?lang=en.
49. Gilding, P. (2020, May 26). It will get darker before the dawn. *Paul Gilding, Cockatoo Chronicles* Retrieved from https://paulgilding.com/2020/05/26/it-will-get-darker-before-the-dawn/.
50. Marshall, G. (2014). *Don't even think about it: Why our brains are wired to ignore climate change.* New York: Bloomsbury.
51. Moore, J. W. (2016). *Anthropocene or capitalocene? Nature, history, and the crisis of capitalism.* Oakland, CA: PM Press.
52. Pew Research Centre. (2019, April 18). A look at how people around the world view climate change. *Pew Research Center.* Retrieved 9 August 2020 from https://www.pewresearch.org/fact-tank/2019/04/18/a-look-at-how-people-around-the-world-view-climate-change/.
53. Lowy Institute. (2019, May 8). 2019 *Lowy Institute poll – Australian attitudes to climate change.* Retrieved 27 July 2020 from https://www.lowyinstitute.org/publications/media-release-2019-lowy-institute-poll-australian-attitudes-climate-change.
54. Rockström, J., Gaffney, O., Rogelj, J., Meinshausen, M., Nakicenovic, N., & Schellnhuber, H. J. (2017). A roadmap for rapid decarbonization. *Science, 355*(6331), 1269–1271. https://doi.org/10.1126/science.aah3443.

55. International Renewable Energy Agency. (2018). *Renewable energy capacity statistics 2018*. Retrieved 10 December 2018 from /newsroom/pressreleases/2018/Apr/Global-Renewable-Generation-Continues-its-Strong-Growth-New-IRENA-Capacity-Data-Shows.
56. International Renewable Energy Agency. (2019). *Renewable power generation costs in 2019* (p. 144). Retrieved 9 August 2020 from https://www.irena.org/-/media/Files/IRENA/Agency/Publication/2020/Jun/IRENA_Power_Generation_Costs_2019.pdf.
57. International Energy Agency. (2018). *World energy outlook 2018*.
58. International Union for Conservation of Nature. (2017). *Deforestation and forest degradation*.
59. Meadows, D. H., Randers, J., Meadows, D. L., & Behrens, W. W. (1974). *The limits to growth: A report for the Club of Rome's Project on the predicament of mankind* (2nd ed.). New York: Universe Books.
60. Fritze, J. G., Blashki, G. A., Burke, S., & Wiseman, J. (2008). Hope, despair and transformation: Climate change and the promotion of mental health and wellbeing. *International Journal of Mental Health Systems, 2*(1), 13. https://doi.org/10.1186/1752-4458-2-13.
61. Hayes, K., Blashki, G., Wiseman, J., Burke, S., & Reifels, L. (2018). Climate change and mental health: Risks, impacts and priority actions. *International Journal of Mental Health Systems, 12*(1). https://doi.org/10.1186/s13033-018-0210-6.
62. Greenberg, P. (2010, May 7). Book review - Eaarth - By Bill McKibben. *The New York Times*. Retrieved 26 November 2018 from https://www.nytimes.com/2010/05/09/books/review/Greenberg-t.html.
63. Head, L. (2016). *Hope and grief in the anthropocene: Re-conceptualising human–nature relations*. New York, NY: Routledge, Taylor & Francis Group. https://doi.org/10.4324/9781315739335.
64. Ellis, A., & Harper, R. (1997). *A guide to rational living* (3rd rev. ed.). Chatsworth, CA: Wilshire.
65. Pilisuk, M., Rowen, J., & Psychologists for Social Responsibility. (2005). *Using psychology to help abolish nuclear weapons; a handbook*. American Psychological Association. https://doi.org/10.1037/e403012005-001.
66. Zinn, H. (2004, September 2). The optimism of uncertainty. *The Nation*. Retrieved from https://www.thenation.com/article/optimism-uncertainty/.
67. Carson, R. (2001). *Silent Spring* (1st ed.). London: Penguin.
68. Carson, R., & Lear, L. J. (1998). *Lost woods: The discovered writing of Rachel Carson*. Boston: Beacon Press.

69. Carson, R., & Kelsh, N. (1998). *The Sense of Wonder* (1st ed.). New York: Harper Collins Publishers.
70. McKibben, B. (2013). *Oil and Honey: The education of an unlikely activist* (1st ed.). New York: Times Books, Henry Holt and Company.
71. Smith, H. (2014, February 5). How 350.org went from "strange kid" to head of the green class. *Grist*. Retrieved 18 December 2018 from https://grist.org/climate-energy/how-350-org-went-from-strange-kid-to-head-of-the-green-class/.
72. Pollak, S. (2014). 350.org's rise from Middlebury College. *The Burlington Free Press*. Retrieved 9 August 2020 from https://www.burlingtonfreepress.com/story/life/green-mountain/2014/11/06/org-middlebury-college/18608541/.
73. Lytton, C. (2013, November 16). Top 10: Climate change campaigns. *The Guardian*. Retrieved 12 January 2020 from https://www.theguardian.com/global-development-professionals-network/2013/nov/15/top-10-climate-change-campaigns.
74. Carrington, D. (2018, September 11). Fossil fuel divestment funds rise to $6tn. *The Guardian*. Retrieved 19 December 2018 from https://www.theguardian.com/environment/2018/sep/10/fossil-fuel-divestment-funds-rise-to-6tn.
75. McKibben, B. (2018, December 16). At last, divestment is hitting the fossil fuel industry where it hurts | Bill McKibben. *The Guardian*. Retrieved 1 October 2019 from https://www.theguardian.com/commentisfree/2018/dec/16/divestment-fossil-fuel-industry-trillions-dollars-investments-carbon.
76. Loeak, M. (2014, October 17). We won't stand by while coal companies destroy our Marshall Islands homes. *The Guardian*. Retrieved 20 December 2018 from https://www.theguardian.com/commentisfree/2014/oct/17/we-wont-stand-by-while-coal-companies-destroy-our-marshall-islands-homes.
77. Loloa, S. (2018). The pacific climate warriors. *Have your sei: Sign the Pacific Climate Warriors Declaration on Climate Change*. Retrieved 12 December 2018 from https://haveyoursei.org.
78. Packard, A. (2014, August 20). What does it mean to be a Warrior? *The Pacific Climate Warriors*. Retrieved 1 February 2019 from http://world.350.org/pacificwarriors/2014/08/20/what-does-it-mean-to-be-a-warrior/.
79. Wecker, K. (2017, November 6). Environmental activists storm open-pit coal mine ahead of climate talks. *DW.COM*. Retrieved 20 December

2018, from https://www.dw.com/en/environmental-activists-storm-open-pit-coal-mine-ahead-of-climate-talks/a-41248945.
80. Miller, K. (2014). Moana: The rising of the sea directed by Peter Rockford Espiritu. *The Contemporary Pacific, 26*(2), 585–587. https://doi.org/10.1353/cp.2014.0033.
81. Lutunutabua, F. (2014, May 30). We're building canoes. *350 Pacific.* Retrieved 2 January 2019, from https://350pacific.org/our-work/building_canoes/.
82. Liberate Tate. (2015). Birthmark. *Liberate Tate.* Retrieved 7 January 2019, from http://www.liberatetate.org.uk/birthmark/.
83. Mahony, E. (2017). Opening spaces of resistance in the corporatized cultural institution: Liberate Tate and the Art Not Oil coalition. *Museum and Society, 15*(2), 126–141.
84. Liberate Tate. (2010, May 1). Where it all began. *Liberate Tate.* Retrieved 1 July 2019 from http://www.liberatetate.org.uk/liberating-tate/about/.
85. Memou, A. (2017). Art, activism and the tate. *Third Text, 31*(5–6), 619–631. https://doi.org/10.1080/09528822.2018.1435086.
86. BP or not BP. (2013, March 18). Our manifesto. *BP or not BP?* Retrieved 7 March 2020 from https://bp-or-not-bp.org/our-manifesto/.
87. Liberate Tate. (n.d.). Tate is liberated from BP by Liberate Tate. Retrieved 3 July 2020, from http://www.liberatetate.org.uk/performances/tate-is-liberated-from-bp-by-liberate-tate/.
88. Savage, S. (2017, March 10). Culture & capital: Liberate tate. *Assemble Papers.* Retrieved 8 January 2019 from https://assemblepapers.com.au/2017/03/20/culture-capital-liberate-tate/.
89. Hockenos, P. (2015, June 22). The history of the Energiewende. *Clean Energy Wire.* Retrieved 7 January 2019, from https://www.cleanenergywire.org/dossiers/history-energiewende.
90. Steinbacher, K., & Pahle, M. (2016). Leadership and the energiewende: German Leadership by Diffusion. *Global Environmental Politics, 16*(4), 70–89. https://doi.org/10.1162/GLEP_a_00377.
91. Morris, C., & Jungjohann, A. (2016). *Energy democracy: Germany's energiewende to renewables.* Palgrave Macmillan. https://doi.org/10.1007/978-3-319-31891-2.
92. Krause, F., Bossel, H., & Müller-Reissmann, K.-F. (1980). *Energie-Wende: Wachstum u. Wohlstand ohne Erdöl u. Uran: e. Alternativ-Bericht d. Öko-Inst., Freiburg.* Frankfurt am Main: S. Fischer.
93. Federal Ministry for the Environment, Nature Conservation and Nuclear Safety. (2010). *The Federal Government's energy concept of 2010 and*

the transformation of the energy system of 2011 (p. 37). Retrieved 7 January 2019 from https://cleanenergyaction.files.wordpress.com/2012/10/german-federal-governments-energy-concept1.pdf.
94. Federal Ministry for the Environment, Nature Conservation and Nuclear Safety. (2016, November 1). Climate action plan 2050. *bmu.de*. Retrieved 7 January 2019, from https://www.bmu.de/PU395-1.
95. Rogge, K. S., & Johnstone, P. (2017). Exploring the role of phase-out policies for low-carbon energy transitions: The case of the German Energiewende. *Energy Research & Social Science, 33*, 128–137. https://doi.org/10.1016/j.erss.2017.10.004.
96. Mathews, J. (2017, October 10). The spectacular success of the German energiewende and what to do next. *EnergyPost.eu*. Retrieved 4 January 2019 from https://energypost.eu/the-spectacular-success-of-the-german-energiewende-and-what-needs-to-be-done-next/.
97. Clean Energy Wire. (2015, April 7). Polls reveal citizens' support for Energiewende. *Clean Energy Wire*. Retrieved 7 January 2019, from https://www.cleanenergywire.org/factsheets/polls-reveal-citizens-support-energiewende.
98. Amelang, S., Egenter, S., Wehrmann, B., & Wettengel, J. (2019, January 8). German politicians and energy experts praise coal exit deal, say real work starts now. *Clean Energy Wire*. Retrieved 9 August 2020, from https://www.cleanenergywire.org/news/german-climate-and-energy-experts-praise-coal-exit-deal-say-real-work-starts-now.
99. Otto, I. M., Donges, J. F., Cremades, R., Bhowmik, A., Hewitt, R. J., Lucht, W., Schellnhuber, H. J. (2020). Social tipping dynamics for stabilizing Earth's climate by 2050. *Proceedings of the National Academy of Sciences, 117*(5), 2354–2365. https://doi.org/10.1073/pnas.1900577117.
100. Klein, N., December 15 2018, 15th, & Comments, 0. (2018, December 15). The sunrise movement and the "green new deal": Climate hope from Trump's America? *Ecologise*. Retrieved 29 December 2018, from https://www.ecologise.in/2018/12/15/the-sunrise-movement-and-the-green-new-deal-climate-hope-in-trumps-america/.
101. Dickinson, T. (2019, January 7). What is the green new deal? *Rolling Stone*. Retrieved 8 January 2019 from https://www.rollingstone.com/politics/politics-news/green-new-deal-explained-775827/.
102. Green, M. (2018, December 3). Ocasio-Cortez: Fighting climate change will be 'the civil rights movement of our generation.' *TheHill*. Text. Retrieved 12 January 2019, from https://thehill.com/policy/energy-env

ironment/419564-ocasio-cortez-fighting-climate-change-will-be-the-civil-rights.
103. Roberts, D. (2018, December 21). The green new deal, explained. *Vox*. Retrieved 12 January 2019, from https://www.vox.com/energy-and-environment/2018/12/21/18144138/green-new-deal-alexandria-ocasio-cortez.
104. Ocasio-Cortez, A. (2018, December 19). Our ultimate end goal isn't a Select Committee. Our goal is to treat Climate Change like the serious, existential threat it is by drafting an ambitious solution on the scale necessary - aka a Green New Deal - to get it done. A weak committee misses the point & endangers people. https://twitter.com/mirandacgreen/status/1075495787050741772 @*aoc*. Tweet. Retrieved from https://twitter.com/aoc/status/1075516273327570945?lang=en.
105. Time, W. D. H. (2018, August 24). Greta Thunberg: "Sweden is not a Role Model." *We Don't Have Time*. Retrieved 2 January 2019 from https://medium.com/@wedonthavetime/greta-thunberg-sweden-is-not-a-role-model-6ce96d6b5f8b.
106. Thunberg, G. (2018). "We have not come here to beg world leaders to care." *Common Dreams*. Retrieved 10 August 2020 from https://www.commondreams.org/news/2018/12/04/we-have-not-come-here-beg-world-leaders-care-15-year-old-greta-thunberg-tells-cop24.
107. O'Shea Carre, H., & Albrecht, M. (2018, October 31). Why we're striking from school over climate change inaction. *The Age*. Retrieved 14 Janaury 2019 from https://www.theage.com.au/national/victoria/why-we-re-striking-from-school-over-climate-change-inaction-20181031-p50d30.html.
108. School Strike 4 Climate. (2019). School Strike 4 Climate overview. *Linked In*. Retrieved 10 August 2020, from https://www.linkedin.com/company/schoolstrike4climate/?originalSubdomain=au.
109. Langford, S. (2018, November 29). Meet the kids who are skipping schools Australia-Wide To Call For Climate Action. *Junkee*. Retrieved 2 January 2019, from https://junkee.com/school-strike-climate-change/184496.
110. Laville, S., & Watts, J. (2019, September 20). Across the globe, millions join biggest climate protest ever. *The Guardian*. Retrieved 29 May 2020 from https://www.theguardian.com/environment/2019/sep/21/across-the-globe-millions-join-biggest-climate-protest-ever.
111. Zhou, N. (2018, November 30). Climate change strike: Thousands of school students protest across Australia. *The Guardian*. Retrieved 2

January 2019 from https://www.theguardian.com/environment/2018/nov/30/climate-change-strike-thousands-of-students-to-join-national-protest.
112. Climate Action Network Australia members. (2020). *Climate action network Australia.* Retrieved 21 November 2020, from https://www.cana.net.au/our_members.
113. Seed Indigenous Youth Climate Network. (2020). Seed. Retrieved 21 November 2020, from https://www.seedmob.org.au/.
114. Grenville, K. (2005). *The secret river.* Edinburgh: Cannongate. Retrieved 10 August 2020 from https://www.amazon.com/Secret-River-Kate-Grenville/dp/1841959146.
115. Wiseman, J. P., & Thomson, D. F. (1996). *Thomson time: Arnhem Land in the 1930s: A photographic essay.* Melbourne, VIC: Museum of Victoria.
116. Allam, L. (2020, January 5). For First Nations people the bushfires bring a particular grief, burning what makes us who we are. *The Guardian.* Retrieved 7 Janaury 2020 from https://www.theguardian.com/commentisfree/2020/jan/06/for-first-nations-people-the-bushfires-bring-a-particular-grief-burning-what-makes-us-who-we-are.
117. Birch, T. (2017, March 3). Climate change, recognition and social place making. *Sydney Review of Books.* Retrieved from https://sydneyreviewofbooks.com/climate-change-recognition-and-caring-for-country/.
118. Rose, D. B. (2004). *Reports from a wild country: Ethics for decolonisation.* Sydney, NSW: UNSW Press.
119. Birch T. (2017) Climate change, recognition and caring for country. *Sydney Review of Books.* Retrieved 8 October 2019, from https://sydneyreviewofbooks.com/climate-change-recognition-and-caring-for-country/.
120. Birch, T. (2018). Recovering a narrative of place. *Griffith Review, 60* First Things First
121. Yunkaporta, T. (2019). *Sand talk: How Indigenous thinking can save the world.* Melbourne, VIC: Text Publishing.
122. Yunkaporta, T. (2019, August 29). Read a Q&A with Tyson Yunkaporta, author of Sand Talk. *The Booktopian.* Retrieved 10 August 2020 from https://www.booktopia.com.au/blog/2019/08/29/read-a-qa-with-tyson-yunkaporta-author-of-sand-talk/.
123. Whyte, K. (2020). Too late for indigenous climate justice: Ecological and relational tipping points. *WIREs Climate Change, 11*(1), e603. https://doi.org/10.1002/wcc.603.

124. Whyte, K. (2018). Settler colonialism, ecology, and environmental injustice. *Environment and Society, 9*(1), 125–144. https://doi.org/10.3167/ares.2018.090109.
125. Kimmerer, R. W. (2015). *Braiding Sweetgrass: Indigenous wisdom, scientific knowledge and the teachings of plants.* Minneapolis, MN: Milkweed Editions.
126. Kimmerer, R. W. (n.d.). Returning the gift. *Gratefulness.org.* Retrieved 10 August 2020, from https://gratefulness.org/resource/returning-the-gift/.
127. Johnston, B. (2002). *The Manitous: The Spiritual World of the Ojibway.* St. Paul, MN: Minnesota Historical Society Press, U.S.
128. Wearden, G. (2019, January 21). David Attenborough tells Davos: 'The Garden of Eden is no more.' *The Guardian.* Retrieved 22 January 2019 from https://www.theguardian.com/tv-and-radio/2019/jan/21/david-attenborough-tells-davos-the-garden-of-eden-is-no-more.
129. Oxford Dictionary. (n.d.). Anthropocene. *Lexico dictionaries | English.* Retrieved 10 August 2020, from https://www.lexico.com/definition/anthropocene.
130. Breakthrough Institute. (2015, April). An ecomodernist manifesto. *Breakthrough Institute.* Retrieved 10 August 2020 from www.ecomodernism.org.
131. Lent, J. (2018, May 21). Steven Pinker's ideas are fatally flawed. These eight graphs show why. *openDemocracy.* Retrieved 10 August 2020 from https://www.opendemocracy.net/en/transformation/steven-pinker-s-ideas-are-fatally-flawed-these-eight-graphs-show-why/.
132. Kingsnorth, P., & Hine, D. (2009). *Uncivilization: The dark mountain manifesto.* Retrieved 11 June 2020 from https://dark-mountain.net/about/manifesto/.
133. Pollitt, J. J. (1972). *Art & experience classical Greece* (1st ed.). Cambridge, New York, Melbourne: Cambridge University Press.
134. Brickhouse, T. C., & Smith, N. D. (2000). *The philosophy of Socrates.* Westview Press.
135. Tillich, P. (1980). *The Courage To Be.* New Haven: Yale University Press.
136. Oreskes, N., & Conway, E. M. (2010). *Merchants of doubt: How a handful of scientists obscured the truth on issues from tobacco smoke to global warming* (1st U.S. ed.). New York: Bloomsbury Press.
137. Brecht, B., Willett, J., Manheim, R., & Fried, E. (1987). *Poems, 1913-1956.* New York: Routledge.

138. Schonfeld, M. (n.d.). *Climate philosophy newsletter vol 4 2010-2011*. Retrieved 3 July 2020, from https://iseethics.files.wordpress.com/2011/06/climate-philosophy-newsletter-vol-4-2010-2011.pdf.
139. Lane, M. S. (2012). *Eco-republic: What the ancients can teach us about ethics, virtue, and sustainable living*. Princeton, NJ: Princeton University Press.
140. *Fairfax climate watch*. (n.d.). Is All Hope Lost? Plato's Cave Meets Climate Change. Retrieved 1 March 2018, from http://www.fairfaxclimatewatch.com/blog/2013/10/is-all-hope-lost-platos-cave-meets-climate-change.html.
141. Aristotle, Thomson, J. A. K., Tredennick, H., & Aristotle. (2004). *The Nicomachean ethics* (Further rev. ed.). London, UK ; New York, NY: Penguin Books.
142. Hulme, M. (2014). Climate change and virtue: An apologetic. *Humanities, 3*(3), 299–312. https://doi.org/10.3390/h3030299.
143. Aristotle. (1991). *The Art of Rhetoric*. London, UK; New York, NY, USA: Penguin Books.
144. Xenophon. (1923). *Xenophon in seven volumes, 4*. Harvard Univesity Press.
145. Lamb, M., & Lane, M. (2016). Aristotle on the ethics of communicating climate change. In C. Heyward & D. Roser (Eds.), *Climate justice in a non-ideal world* (pp. 229–254). Oxford University Press. https://doi.org/10.1093/acprof:oso/9780198744047.003.0012.
146. Epictetus, & Dobbin, R. F. (2007). *Discourses. Book 1*. Oxford: Clarendon Press.
147. Robertson, D., & Codd, T. (2019). Stoic philosophy as a cognitive-behavioral therapy. *The Behavior Therapist, 42*(2). Retrieved 12 August 2020 from https://medium.com/stoicism-philosophy-as-a-way-of-life/stoic-philosophy-as-a-cognitive-behavioral-therapy-597fbeba786a.
148. Marcus Aurelius, & Clay, D. (2006). *Meditations* (M. Hammond, Trans.). London, New York, Toronto: Penguin Books.
149. Kant, I. (2009). *Answer the question: What is enlightenment?* (T. Books, Ed.). London, UK: Penguin.
150. Israel, J. I. (2002). *Radical enlightenment: Philosophy and the making of modernity 1650-1750*. Oxford: Oxford University Press.
151. Descartes, R. (1641). *Meditations on first philosophy in which the existence of god and the immortality of the soul are demonstrated*.
152. Locke, J. (1937). *Treatise of civil government and a letter concerning toleration*. New York: Appleton Century.

153. Kant, I., & Pluhar, W. S. (1987). *Critique of judgement.* Hackett Publishing.
154. Irwin, R. (Ed.). (2010). *Climate change and philosophy: Transformational possibilities.* London; New York, NY: Continuum.
155. Lent, J. (2017, July 17). Our values will decide our destiny. *Medium.* Retrieved 12 August 2020 from https://medium.com/@jeremylent/our-values-will-decide-our-destiny-b0bc3c784d14.
156. Hegel, G., & Miller, A. (1977). *Phenomenology of Spirit.* Oxford UK: Clarendon Press.
157. Nietzsche cited in Duignan, B. (ed). (2011). *Modern philosophy: From 1500CE to the present.* New York, NY: Brittanica Educational Publishing.
158. Malik, K. (June 2013). Seeing reason: Jonathan Israel's radical vision. *New Humanist.* Retrieved 24 May 2020 from https://newhumanist.org.uk/4194/seeing-reason-jonathan-israels-radical-vision.
159. Wordsworth, W. (1798). Lines composed a few miles above Tintern Abbey, on revisiting the banks of the Wye During a tour, July 13, 1798. *The Guardian.* Retrieved 17 July 2020, from https://www.theguardian.com/books/2010/jan/26/william-wordsworth-lines-composed-a-few-miles-above-tintern-abbey.
160. Keats, J. (2009). *Selected letters of John Keats.* Harvard University Press.
161. Wollestonecraft, M. (1796). *A vindication of the rights of woman: With strictures on political and moral subjects.* London, UK: J. Johnson.
162. Wysession, M. (2018, October 26). Frankenstein meets climate change: Monsters of our own making. *Common Reader.* Retrieved 3 March 2019 from https://commonreader.wustl.edu/c/frankenstein-meets-climate-change-monsters-of-our-own-making/.
163. Wood, G. D. (2015, December 31). The volcano that shrouded the earth and gave birth to a monster. *Nautilus.* Retrieved 12 August 2020, from http://nautil.us/issue/31/stress/the-volcano-that-shrouded-the-earth-and-gave-birth-to-a-monster.
164. Engels, F. (1970). *Dialectics of Nature,* Moscow: Progress Press
165. Foster, J. B. (1999). Marx's theory of metabolic rift: Classical foundations for environmental sociology. *American Journal of Sociology, 105*(2), 366–405. https://doi.org/10.1086/210315.
166. Dawson, A. (2016). *Extinction: A radical history.* New York, NY: OR Books.

167. Clausen, R., & Clark, B. (2005). The metabolic rift and marine ecology: An analysis of the ocean crisis within capitalist production. *Organization & Environment, 18*(4), 422–444. https://doi.org/10.1177/1086026605281187.
168. Foster, J. B. (2017, November). The long ecological revolution. *Monthly Review*. Retrieved 12 August 2020 from https://monthlyreview.org/2017/11/01/the-long-ecological-revolution/.
169. Owen, W., & Stallworthy, J. (2004). *Wilfred Owen: Poems*. London: Faber and Faber.
170. Woolf, V. (1981). *To the lighthouse*. New York: Harvest.
171. Frankl, V. E., Winslade, W. J., & Kushner, H. S. (2006). *Man's search for meaning* (1 ed.). Boston: Beacon Press.
172. Auden, W. H., & Mendelson, E. (1991). *Collected poems* (1st Vintage International ed.). New York: Vintage International, Vintage Books.
173. Adorno, T. W., & Ashton, E. B. (1973). *Negative dialectics*. New York, NY: Seabury Press.
174. Zuidervaart, L. (2015). Theodor W. Adorno. In E. N. Zalta (Ed.), *The Stanford encyclopedia of philosophy* (Winter 2015). Metaphysics Research Lab, Stanford University. Retrieved from https://plato.stanford.edu/archives/win2015/entries/adorno/.
175. Jeffries, S. (2016). *Grand hotel Abyss: The lives of the Frankfurt School*. London; New York: Verso, an imprint of New Left Books.
176. Richardson, J. (2012). *Heidegger*. New York: Routledge.
177. Heidegger, M., & Krell, D. F. (1993). *Basic writings: From Being and time (1927) to the task of thinking (1964)* (Rev. and expanded ed.). San Francisco, CA: HarperSanFrancisco.
178. Gill, G. (2002). Landscape as symbolic form: Remembering thick place in deep time. *Critical Horizons, 3*(2), 177–199. https://doi.org/10.1163/156851602760586644.
179. May, R. (2007). *Love and will*. New York: W.W. Norton.
180. Arendt, H. (1973). *The origins of totalitarianism* (New ed.). New York: Harcourt Brace Jovanovich.
181. Stephenson, W. (n.d.). Learning to live in the dark: Reading arendt in the time of climate change. *Los Angeles Review of Books*. Retrieved 5 March 2019, from https://lareviewofbooks.org/article/learning-to-live-in-the-dark-reading-arendt-in-the-time-of-climate-change/.
182. Arendt, H. (1971). Thinking and moral consideration: A lecture. *Social Research, 38*(3), 417–446.

183. Arendt, H. (1995). *Men in dark times*. San Diego, CA: Harcourt, Brace & Company.
184. Hunt, E. (2019, 12 March) BirthStrikers: Meet the women who refuse to have children until climate change ends. *The Guardian*. Retrieved 8 April 2019, from https://www.theguardian.com/lifeandstyle/2019/mar/12/birthstrikers-meet-the-women-who-refuse-to-have-children-until-climate-change-ends.
185. The Citizens' Assembly of Ireland. (n.d.). Retrieved 26 November 2020, from https://www.citizensassembly.ie/en/.
186. Gibbons, J. (2019). Ireland must listen to Citizens' Assembly on climate change. *The Irish Times*. Retrieved 25 March 2019, from https://www.irishtimes.com/opinion/ireland-must-listen-to-citizens-assembly-on-climate-change-1.3515211.
187. Habermas, J., MacCarthy, T., & Habermas, J. (2007). *Reason and the rationalization of society* (Nachdr.). Boston: Beacon.
188. Habermas, J., & Rehg, W. (2010). *Contributions to a discourse theory of law and democracy* (Reprinted.). Cambridge: Polity Press.
189. Brulle, R. J. (2002). Habermas and green political thought: Two roads converging. *Environmental Politics, 11*(4), 1–20. https://doi.org/10.1080/714000651.
190. Eckersley, R. (1999). The discourse ethic and the problem of representing nature. *Environmental Politics, 8*(2), 24–49. https://doi.org/10.1080/09644019908414460.
191. Frankl, V. (2006). *Man's search for meaning*. Boston: Beacon Press.
192. Bonhoeffer, D., & Eberhard, B. (1971). *Letters and papers from prison*. New York: Macmillan.
193. Ocasio-Cortez, A. The democrat who challenged her party's establishment — And won. *The Washington Post*. Retrieved 17 July 2020, from https://www.washingtonpost.com/news/powerpost/wp/2018/06/27/alexandria-ocasio-cortez-the-democrat-who-challenged-her-partys-establishment-and-won/.
194. Sartre, J.-P., Kulka, J., Elkaïm-Sartre, A., Cohen-Solal, A. & Macomber, C. (2007). *Existentialism is a humanism =: (L'Existentialisme est un humanisme)*. New Haven: Yale University Press.
195. Beauvoir, S. de, & Frechtman, B. (2015). *The ethics of ambiguity*. New York: Philosophical Library: Distributed by Open Road Integrated Media.
196. Klein, N. (2016, June 2). Let them drown, the violence of othering in a warming world. *London Review of Books*, 11–14.
197. Camus, A. (2000). *The myth of sysiphus*. London: Penguin.

198. Camus, A., & Thody, P. (1970). *Lyrical and critical essays*. New York: Vintage Books.
199. Camus, A. (1991). *The plague* (1st Vintage international ed.). New York: Vintage Books.
200. Hamilton, C. (2011). *Requiem for a species: Why we resist the truth about climate change*. Sydney, NSW: Allen and Unwin.
201. Keimowitz, A. S. (2018, March 9). I felt despair about climate change—Until a brush with death changed my mind. *Slate Magazine*. Retrieved 12 March 2018, from https://slate.com/technology/2018/03/an-environmental-professor-on-learning-to-cope-with-climate-change.html.
202. Monbiot, G. (2009, August 18). Should we seek to save industrial civilisation? *George Monbiot*. Retrieved 8 April 2019, from https://www.monbiot.com/2009/08/18/should-we-seek-to-save-industrial-civilisation/.
203. Francis. (2015). *Laudato Si*. Vatican. http://www.vatican.va/content/vatican/en.html.
204. Jenkins, W., Tucker, M. E., & Grim, J. (Eds.). (2017). *Routledge handbook of religion and ecology*. London: New York: Routledge/Taylor & Francis Group.
205. White, L. (1967). The historical roots of our ecologic crisis. *Science, New Series, 155*(3767), 1203–1207.
206. Saint Francis. (n.d.). *The canticle of brother sun and sister moon*. Retrieved 13 August 2020, from http://www2.webster.edu/~barrettb/canticle.htm.
207. Antal, J., & McKibben, B. (2018). *Climate church, climate world: How people of faith must work for change*. Lanham: Rowman & Littlefield Publishers.
208. Hayhoe, K., & Farley, A. (2009). *A climate for change: Global warming facts for faith-based decisions* (1st ed.). New York: FaithWords.
209. Watts, J. (2019, January 6). Katharine Hayhoe: A thermometer is not liberal or conservative. *The Observer*. Retrieved 13 August 2019 from https://www.theguardian.com/science/2019/jan/06/katharine-hayhoe-interview-climate-change-scientist-crisis-hope.
210. Smith, S. (2016, May). Unfriendly climate. *Texas Monthly*. Retrieved 13 August 2019 from https://www.texasmonthly.com/articles/katharine-hayhoe-lubbock-climate-change-evangelist/.
211. Hanley, S. (2019, January 7). Katharine Hayhoe talks about why climate change is real & why she still has hope. *CleanTechnica*. Retrieved 16 June 2020 from https://cleantechnica.com/2019/01/07/katharine-hayhoe-talks-about-why-climate-change-is-real-why-she-still-has-hope/.

212. Q&A with Climate Scientist Katharine Hayhoe. (2019, January 28). *Rare*. Retrieved 16 June 2020 from https://rare.org/story/lets-talk-solutions/.
213. Waskow, A. (2001, August 9). Between the fires: A litany of grief & hope. *The Shalom Center*. Retrieved 13 August 2020, from https://theshalomcenter.org/node/276.
214. Tirosh-Samuelson, H. (2001). Nature in the sources of Judaism. *Daedalus, 130*(4), 99–124.
215. Shapiro, R. M. (Ed.). (2006). *Ethics of the sages: Pirke Avot-- annotated & explained*. Woodstock, Vt: SkyLight Paths.
216. The Shalom Center. (2018, July 11). A flaming fire, consuming everything - Tisha B'Av in a time of climate crisis. *The Shalom Center*. Retrieved 11 July 2019, from https://theshalomcenter.org/content/flaming-fire-consuming-everything-tisha-bav-time-climate-crisis.
217. Wisdom In Nature. (n.d.). Wisdom In nature: Islamic ecology, Retrieved 18 June 2020, from https://www.wisdominnature.org.
218. United Nations Environment Program. (2018, June 21). How Islam can represent a model for environmental stewardship. *UNEP*. Retrieved 11 July 2019, from http://www.unenvironment.org/news-and-stories/story/how-islam-can-represent-model-environmental-stewardship.
219. United Nations Framework Convention on Climate Change. (2015, August 18). Islamic Declaration on Climate Change. *UNFCCC*. Retrieved 22 August 2019, from https://unfccc.int/news/islamic-declaration-on-climate-change.
220. Islamic Climate Change Declaration. (2015) Retrieved from https://www.ifees.org.uk/about/islamic-declaration-on-global-climate-change/.
221. Skrimshire, S. (2019, May 12). Extinction rebellion and the new visibility of religious protest. *openDemocracy*. Retrieved from https://www.opendemocracy.net/en/transformation/extinction-rebellion-and-new-visibility-religious-protest/.
222. Wisdom in Nature. (2019, September 17). XR Faith Bridge 2019 (London): Muslims XR Statement & Invite. Retrieved 18 June 2020, from https://www.wisdominnature.org/blog/faith-bridge-muslims-xr.
223. Welch, B. (2019, July 8). Extinction rebellion: Activism rooted in "Jewish values." *The Jewish Chronicle*. Retrieved 14 August 2020 from https://www.thejc.com/news/uk/extinction-rebellion-jews-xr-jewish-values-climate-change-catastrophe-1.486255.
224. Rolles, G. (2020). I got arrested for you. *ARRCC*. Retrieved 18 June 2020 from https://www.arrcc.org.au/i_got_arrested_for_you.

225. Global Buddhist Climate Change Collective. (2015, October 29). Buddhist climate change statement to world leaders. *GBCCC.* Retrieved 14 August 2020 from http://gbccc.org/.
226. Lion's Roar. (2016, January 25). Christiana Figueres cites Thich Nhat Hanh's influence in Paris climate talks. *Lion's Roar.* Retrieved 14 August 2020 from https://www.lionsroar.com/christiana-figueres-cites-thich-nhat-hanhs-influence-in-paris-climate-talks/.
227. Thich Nhat Hanh's statement on Climate Change for the United Nations. (2015, July 2). *Plum Village.* Retrieved 2 May 2018, from https://plumvillage.org/letters-from-thay/thich-nhat-hanhs-statement-on-climate-change-for-unfccc/.
228. Nhất Hạnh. (1991). *Peace is every step: The path of mindfulness in everyday life.* (A. Kotler, Ed.). New York, NY: Bantam Books.
229. Macy, J. (n.d.). *Joannamacy.net.* Retrieved 14 August 2020 from http://www.joannamacy.net/html/engaged.html.
230. Gregoire, C. (2014, May 19). Jack Kornfield on gratitude and mindfulness. *Greater Good Magazine.* Retrieved 14 August 2020, from https://greatergood.berkeley.edu/article/item/jack_kornfield_on_gratitude_and_mindfulness.
231. Niehaus, D. (n.d.). To love the earth: The ecological vision in Gary Snyder's ecopoetry. Retrieved 14 August 2020 from https://www.academia.edu/2588312/To_Love_the_Earth_The_Ecological_Vision_in_Gary_Snyders_Ecopoetry.
232. Hallisey, C. (Ed.). (2015). *Therigatha: Poems of the first Buddhist women.* Cambridge, MA: Harvard University Press.
233. Hanshan, Shide, Wang, F., & Seaton, J. P. (2019). *Cold mountain poems: Zen poems of Han Shan, Shih Te, and Wang Fan-Chih.*
234. Rand, A., & Peikoff, L. (1993). *The Fountainhead.* New York: Signet.
235. Freedland, J. (2017, April 11). The new age of Ayn Rand: How she won over Trump and Silicon Valley. *The Guardian Australia.* Retrieved 14 August 2020 from https://www.theguardian.com/books/2017/apr/10/new-age-ayn-rand-conquered-trump-white-house-silicon-valley.
236. Rand, A., & Bastide-Foltz, S. (2017). *La grève: Atlas shrugged.*
237. Ringu Tulku, & Fuchs, R. (2005). *Daring steps toward fearlessness: The three vehicles of Buddhism.* Ithaca, NY: Snow Lion Publications.
238. Loy, D. (2019, September 18). Can Buddhism meet the climate crisis? *Lion's Roar.* Retrieved 14 August 2020 from https://www.lionsroar.com/can-buddhism-meet-the-climate-crisis/.

239. Loy, D. (2014, May 12). Listening to the Buddha: How greed, ill-will and delusion are poisoning our institutions. *openDemocracy*. Retrieved 14 August 2020 from https://www.opendemocracy.net/en/transformation/listening-to-buddha-how-greed-illwill-and-delusion-are-poisoning-our-institut/.
240. Aitken, R. (2003). Giving full circle. *Tricycle: The Buddhist Review*. Retrieved 15 April 2019 from https://tricycle.org/magazine/giving-full-circle/.
241. Brown, C. (2017). *Buddhist economics: An enlightened approach to the dismal science*. New York: Bloomsbury Press.
242. Schumacher, E. F. (1993). *Small is beautiful: Economics as if people mattered*. London: Vintage Books.
243. Thinley, J. (2012, June 20). Statement by the Prime Ministe of Bhutan to the United Nations Conference on Sustainable Development (Rio+20). Retrieved 23 December 2020 from https://sustainabledevelopment.un.org/content/documents/16693bhutan.pdf.
244. Rayworth, K. (2017). *Doughnut Economics: Seven ways to think like a 21st-century economist*. Cornerstone Digital.
245. Rayworth, K. (2013). What on earth is the doughnut? Retrieved 15 August 2020 from https://www.kateraworth.com/doughnut/.
246. Rockström, J., Steffen, W., Noone, K., Persson, Å., Chapin, F. S. I., Lambin, E., Foley, J. (2009). Planetary boundaries: Exploring the safe operating space for humanity. *Ecology and Society, 14*(2). https://doi.org/10.5751/ES-03180-140232.
247. Boffey, D. (2020, April 8). Amsterdam to embrace "doughnut" model to mend post-coronavirus economy. *The Guardian*. Retrieved 23 December 2020 from https://www.theguardian.com/world/2020/apr/08/amsterdam-doughnut-model-mend-post-coronavirus-economy.
248. United Nations. (2020). The recovery from the COVID-19 crisis must lead to a different economy. *United Nations*. Retrieved 23 December 2020, from https://www.un.org/en/un-coronavirus-communications-team/launch-report-socio-economic-impacts-covid-19.
249. Halifax, R. J. (2019, January 1). Wise hope in social engagement by Roshi Joan Halifax - Part 1. *Upaya Zen Center*. Retrieved 24 May 2020 from https://www.upaya.org/2018/12/wise-hope-in-social-engagement-by-roshi-joan-halifax-part-1/.
250. Halifax, R. J. (2018, December 19). Roshi Joan Halifax delivers contemplation by design keynote, "The Strange and Necessary Case for Hope." Retrieved 15 August 2020 from https://religiouslife.stanford.edu/

news/roshi-joan-halifax-delivers-contemplation-design-keynote-strange-and-necessary-case-hope.
251. Laozi, & Mitchell, S. (2006). *Tao te ching: A new English version.* New York: HarperCollins.
252. Watts, A., & Huang, A. C.-L. (1975). *Tao: The watercourse way.* New York, NY: Pantheon Books.
253. Liu, A., & Major, J. S. (Eds.). (2012). *The essential Huainanzi.* New York: Columbia University Press.
254. Xia, C., & Schönfeld, M. (2011). A daoist response to climate change. *Journal of Global Ethics, 7*(2), 195–203. https://doi.org/10.1080/17449626.2011.590279.
255. Chinese Daoist Association. (n.d.). *Daoist faith statement on the environment.* Retrieved 24 May 2020 from https://www.interfaithsustain.com/daoist-faith-statement-on-the-environment/.
256. Ming Wang. (1960). *Taiping jing hejiao.* Beijing: Zhongua Publishing.
257. Kasoff, I. E. (2002). *The thought of Chang Tsai (1020-1077)* (1. paperback ed.). Cambridge: Cambridge University Press.
258. Alliance of Religions and Conservation. (2013). *What does Confucianism teach about ecology?* Retrieved 15 August 2020, from http://www.arcworld.org/faiths.asp?pageID=182.
259. Mingming, W. (2012). All under heaven (tianxia): Cosmological perspectives and political ontologies in pre-modern China. HAU: *Journal of Ethnographic Theory, 2*(1), 337–383. https://doi.org/10.14318/hau2.1.015.
260. Han, S.-J., Shim, Y.-H., & Park, Y.-D. (2016). Cosmopolitan sociology and confucian worldview: Beck's theory in East Asia. *Theory, Culture & Society, 33*(7–8), 281–290. https://doi.org/10.1177/0263276416672535.
261. Sang-Jin Han Young-Do Park. (n.d.). *Another cosmopolitanism: A critical reconstruction of neo-confucian conception of Tianxiaweigong in the age of global risks.* Retrieved 15 August 2020, from http://gqhfund.jlu.edu.cn/info/1010/1113.htm.
262. Donne, J. (n.d.). *Devotions upon emergent occasions.* Ann Arbour: University of Michigan Press.
263. Atwood, M. (2012, December 7). Margaret Atwood: Rachel Carson's silent spring, 50 years on. *The Guardian.* Retrieved 27 September 2019 from https://www.theguardian.com/books/2012/dec/07/why-rachel-carson-is-a-saint.
264. Carson, R. (2018). *The Sea Around Us.* New York: Oxford University Press.

265. Willis, A. J. (1997). The ecosystem: An evolving concept viewed historically. *Functional Ecology, 11*(2), 268–271. https://doi.org/10.1111/j.1365-2435.1997.00081.x.
266. Trudgill, S. (2007). Tansley, A.G. 1935: The use and abuse of vegetational concepts and terms. Ecology 16, 284—307. *Progress in Physical Geography: Earth and Environment, 31*(5), 517–522. https://doi.org/10.1177/0309133307083297.
267. Elton, C. (1958). *The ecology of invasions by animals and plants.* London: Methuen.
268. Thomas, L. (2020, June 8). Why every environmentalist should be anti-racist. *Vogue.* Retrieved 25 June 2020, from https://www.vogue.com/article/why-every-environmentalist-should-be-anti-racist.
269. Harrison, R. (2013). *Animal machines.* London, UK: Vincent Stuart.
270. van Boekel, J. (n.d.). An ecology of mind - The Gregory Bateson documentary. Retrieved 15 August 2020, from http://www.natureareteducation.org/AnEcologyOfMind.htm.
271. Bateson, G. (1973). *Steps to an ecology of mind.* Paladin.
272. Bateson, G. (2015). *An ecology of mind, The Gregory Bateson documentary.* Retrieved 15 September 2019 from https://www.youtube.com/watch?time_continue=133&v=AqiHJG2wtPI.
273. Bowers, C. A. (2012). The relevance of Gregory Bateson's ideas to understanding the cultural roots of the ecological crisis, 39, 7–38.
274. Meadows, D. (2012, January 19). Speak the truth. *The Donella Meadows Project.* Retrieved 15 August 2020, from http://donellameadows.org/speak-the-truth/.
275. Meadows, D. H., & Wright, D. (2009). *Thinking in Systems: A primer.* London, UK: Earthscan.
276. Meadows, D. (2001, February 2). Every five years…. *The Donella Meadows Project.* Retrieved 15 August 2020 from http://donellameadows.org/every-five-years/.
277. Naess, A. (1973). The shallow and the deep, long-range ecology movement. A summary. *Inquiry, 16*(1–4), 95–100. https://doi.org/10.1080/00201747308601682.
278. Baard. (2015). Managing climate change: A view from deep ecology. *Ethics and the Environment, 20*(1), 23. https://doi.org/10.2979/ethicsenviro.20.1.23.
279. Næss, A., Glasser, H., & Drengson, A. R. (2005). *The selected works of Arne Naess. Volume 1 Volume 1.* Dordrecht: Springer. Retrieved 8 July 2020 from http://site.ebrary.com/id/10217931.

280. Naess, A., Drengson, A., & Inouye, Y. (1995). Self-realization: An ecological approach to being in the world. *The Deep Ecology Movement*. Berkeley CA: North Atlantic Books.
281. Naess, A., Drengson, A., & Devall, B. (2008). *The Ecology of Wisdom*. Berkeley, CA: Counterpoint Press.
282. van Boekel, J. (1995, June). Interview with Norwegian eco-philospher Arne Naess. Retrieved 15 August 2020 from http://www.natureareducation.org/Interview_Arne_Naess_1995.pdf.
283. Blasdel, A. (2017, June 15). A reckoning for our species: The philosopher prophet of the Anthropocene. *The Guardian*. Retrieved 24 May 2020 from https://www.theguardian.com/world/2017/jun/15/timothy-morton-anthropocene-philosopher.
284. Agresta, M. (2018, April 12). Timothy Morton is Houston's own Catastrophe Guru. *Texas Monthly*. Retrieved 24 May 2020 from https://www.texasmonthly.com/the-culture/timothy-morton-catastrophe-houston/.
285. Morton, T. (2012). *Ecological thought* (1st ed.). Cambridge, MA: Harvard University Press.
286. Morton, T. (2018). *Being ecological*. London: Penguin.
287. Dostoyevsky, F. (2012). *The Brothers Karamazov*. Dover.
288. Griffin, D. P. (2017). *CDP carbon majors report 2017, 16*.
289. Smith, P. D. (2018, January 20). Being Ecological by Timothy Morton review – A playfully serious look at the environment. *The Guardian*. Retrieved 16 August 2020 from https://www.theguardian.com/books/2018/jan/20/being-ecological-timothy-morton-review.
290. Selberg, C. A. (1993). *The Revised Foil*. Ashland, OR: Spotted Dog Pr.
291. Triolo, N. (2018, September 2). Four questions for the author: Timothy Morton. *Orion Magazine*. Retrieved 17 January 2020 from https://orionmagazine.org/2018/09/four-questions-for-the-author-timothy-morton-being-ecological/.
292. Schmidgen, H. (2014). *Bruno Latour in pieces: An intellectual biography*. Fordham University Press.
293. Latour, B. (2018). *Down to earth: Politics in the new climatic regime*. Cambridge: Polity Press.
294. Hall, S. (2015, October 26). Exxon knew about climate change almost 40 years ago. *Scientific American*. Retrieved 8 December 2020, from https://www.scientificamerican.com/article/exxon-knew-about-climate-change-almost-40-years-ago/.
295. Plumwood, V. (2003). *Feminism and the Mastery of Nature*. London; New York: Routledge.

296. Shiva, V. (2019, January 5). Eco-activist Vandana Shiva finds reasons for hope in the climate crisis. *CBC Radio*. Retrieved 28 December 2020, from https://www.cbc.ca/radio/tapestry/finding-hope-in-the-climate-crisis-1.4950656/eco-activist-vandana-shiva-finds-reasons-for-hope-in-the-climate-crisis-1.4950685.
297. Hopkins, R. (2019). *From What is to What If: Unleashing the Power of Imagination to Create the Future we Want*. White River Junction: Chelsea Green Publishing.
298. Fisher, M. (2009). *Capitalist Realism: Is there no Alternative?* Winchester: O Books.
299. Hajer, M., & Versteeg, W. (2019). Imagining the post-fossil city: Why is it so difficult to think of new possible worlds? *Territory, Politics, Governance*, 7(2), 122–134. https://doi.org/10.1080/21622671.2018.1510339.
300. Monbiot, G. (2017, September 9). George Monbiot: How do we get out of this mess? *The Guardian*. Retrieved 5 May 2020 from https://www.theguardian.com/books/2017/sep/09/george-monbiot-how-de-we-get-out-of-this-mess.
301. Levitas, R. (2017). *Where there is no vision the people perish: A utopian ethic for a transformed future*. Centre for the Understanding of Sustainable Prosperity. Retrieved from http://www.cusp.ac.uk/wp-content/uploads/05-Ruth-Levitas-Essay-online.pdf.
302. Mielville, C. (2018, March 2). The limits of utopia. Retrieved from https://climateandcapitalism.com/2018/03/02/china-mieville-the-limits-of-utopia/.
303. Wright, E. O. (2010). *Envisioning Real Utopias*. London; New York: Verso.
304. Friedman, M. (2009). *Capitalism and Freedom: Fortieth anniversary edition*. University of Chicago Press.
305. Klein, N. (2020, March 16). Coronavirus capitalism — And how to beat it. *The Intercept*. Retrieved 16 August 2020 from https://theintercept.com/2020/03/16/coronavirus-capitalism/.
306. Sawin, E. (2020, April 8). COVID-19 is many things. *Elizabeth Sawin on Twitter*. Retrieved 16 August 2020, from https://twitter.com/bethsawin/status/1247852845610778624.
307. Robinson, K. (2011). Remarks on utopia in the age of climate change. In A. Milner, S. Sellars, & V. Burgman (Eds.), *Changing the climate: Utopia, dystopia and catastrophe*. Arena Publications.

308. Robinson, K. (2020, May 1). The coronavirus is rewriting our imaginations. *The New Yorker.* Retrieved 15 August 2020 from https://www.newyorker.com/culture/annals-of-inquiry/the-coronavirus-and-our-future.
309. Figueres, C., & Rivett-Carnac, T. (2020). *The Future we Choose.* London: Penguin Random House. Retrieved 16 August 2020 from https://www.penguinrandomhouse.com/books/623543/the-future-we-choose-by-christiana-figueres-and-tom-rivett-carnac/.
310. Bastide, X. (2020). Imagine the future. *TED-Ed.* Retrieved 1 July 2020, from https://ed.ted.com/best_of_web/kpN3w89R.
311. Shue, H. (2014). *Climate Justice: Vulnerability and Protection* (1st ed.). Oxford: Oxford University Press.
312. Bali Principles of Climate Justice. (2002, August 28). Retrieved 1 May 2020, from http://www.indiaresource.org/issues/energycc/2003/baliprinciples.html.
313. World People's Conference on Climate Change and the Rights of Mother Earth Peoples Agreement. (2010, January 5). Retrieved 24 May 2020 from https://pwccc.wordpress.com/support/.
314. Schlosberg, D. (2019, June 25). Can the idea of justice extend to the nonhuman world? *ABC Religion & Ethics.* Opinion, Australian Broadcasting Corporation. Retrieved 30 April 2020 from https://www.abc.net.au/religion/an-ethic-of-ecological-justice-for-the-anthropocene/11246010.
315. Wahl, D. C. (2016). *Designing Regenerative Cultures.* Axminster, UK: Triarchy Press.
316. Regeneration International. (2020). *Regeneration international.* Retrieved 28 December 2020, from https://regenerationinternational.org/.
317. Haraway, D. J. (2016). *Staying with the Trouble: Making Kin in the Chthulucene.* Duke University Press Books.
318. Solnit, R. (2010). *A Paradise Built in Hell: The extraordinary communities that arise in disaster.* New York: Penguin Books.
319. Jemisin, N. K. (2018). *The Broken Earth Trilogy.* New York, NY: Orbit.
320. Iles, A. (2019). Repairing the broken earth: N.K. Jemisin on race and environment in transitions. *Elem Sci Anth, 7*(1), 26. https://doi.org/10.1525/elementa.364.
321. Robinson, K. S. (2020). *The Ministry for the Future.* (1st ed.). New York, NY: Orbit.
322. O'Keefe, D. (2020, October 22). Imagining the end of capitalism with Kim Stanley Robinson. *Jacobin.* Retrieved 22 December

2020, from https://jacobinmag.com/2020/10/kim-stanley-robinson-ministry-future-science-fiction.
323. Monbiot, G. (2015, December 12). Grand promises of Paris climate deal undermined by squalid retrenchments. *The Guardian*. Retrieved 17 July 2020 from https://www.theguardian.com/environment/georgemonbiot/2015/dec/12/paris-climate-deal-governments-fossil-fuels.
324. German Advisory Council on Global Change (Ed.). (2011). *World in Transition: A Social Contract for Sustainability.* Berlin: German Advisory Council on Global Change.

Index

A

adaptation 5, 26, 69, 71, 72, 98, 132, 206, 221
Adorno, Theodor 14, 101–103
Al-Jayyousi 138, 139
Allam, Loreena 62
altruism 15, 116, 158, 208
Anderson, Kevin 26
Angelou, Maya 9
Antal, Jim 128–130
Antarctic 25, 218
Anthropocene 78, 97, 181, 183–185, 188
Apatu, Teuila 17, 212
Arendt, Hannah 14, 105–107
Aristotle 13, 81, 85–87
Attenborough, David 13, 24, 77, 78
Atwood, Margaret 168
Auden, W.H. 100, 101
Aurelius, Marcus 87, 88

B

Baldwin, James 20
Bali Principles of Climate Justice 204, 205, 223, 235
Bastida, Xie 203
Bastida, Xiye 202
Bateson, Gregory 16, 167, 171–173
beauty 15, 16, 18, 37–40, 46, 60, 75, 93, 99, 100, 111, 112, 126, 128–130, 139, 150, 152, 157, 173, 183, 193, 230, 231
Bellamy Foster, John 95–97
Bhutan 155
Biden, Joe 26, 56
Birch, Tony 13, 63–65
Bird Rose, Deborah 63
Black Deaths in Custody 76
Black Lives Matter 76, 170, 230
Blake, William 93
blockchain 221

Boenhoeffer, Dietrich 112
Bostrom, Nick 26
Breakthrough Institute 14, 78, 79
Brecht, Bertolt 82
Buddhist 8, 9, 16, 128, 144–155, 158, 159, 162, 165, 179, 182

C

Camus, Albert 15, 117–119
capitalist 14, 30, 89, 96, 97, 102, 103, 120, 239
capitalist realism 196
Capitolocene 97
care 6, 15, 18, 57, 63, 65–67, 74, 104, 105, 107, 112, 121, 123, 126, 128–131, 138, 149, 155, 157, 160, 183, 189, 193, 200, 204, 206, 207, 209, 229–231
caring for country 13, 65
Carson, Rachael 12, 16, 38–41, 125, 167, 168, 170, 171, 173
Christian 11, 15, 123–125, 129, 131, 132, 137, 143, 144, 169
Christians for Climate Action 143, 144
Citizen's assembly 107, 108
Climate Action Tracker 24, 25
climate crisis 6, 14, 33, 55, 59, 67, 74, 75, 78, 80, 92, 97, 107, 124, 125, 129, 135, 145, 153, 156, 159, 160, 185
climate denial 27, 28, 82
climate emergency 15, 20, 38, 55, 59, 65, 70, 81, 87, 88, 120, 123, 130, 133, 139, 141, 144, 164, 169, 184, 192, 195, 197, 209, 211, 217, 219
climate grief 6, 10, 33, 34, 181

climate justice 12, 15, 32, 48, 60, 63, 65, 66, 69, 71, 98, 128, 129, 131, 135, 144, 193, 204, 206, 220, 229, 235–238
climate science 2, 21, 22, 24, 28, 33, 82, 130, 131, 152, 172, 186, 215
CO_2 emissions 4, 22, 31, 51, 72, 186, 203, 204, 216
Cognitive Behavioural Therapy (CBT) 34, 87, 88
collective action 5, 12, 20, 34, 37–39, 41, 49, 50, 65, 106, 130, 230, 231, 240
colonial invasion 63, 64, 70
communicative action 109–111, 164
compassion 3, 11, 16–18, 34, 37, 39, 128, 130, 141, 142, 145, 148, 153, 154, 157, 158, 162, 165, 204, 208, 210, 229–231
Confucius 162
consumption 4, 11, 31, 32, 41, 75, 85, 114, 128, 130, 140, 142, 146, 175, 177–179, 186, 192, 196, 205, 206, 213, 221–223, 236
COP 21 47, 213
courage 2, 3, 6, 7, 9–12, 15–17, 20, 37, 39, 44, 75, 80–82, 89, 101, 106, 111, 112, 117, 118, 121, 123, 124, 128, 129, 131, 185, 193, 208, 210, 227
COVID-19 5, 22, 112, 156, 200, 201, 211, 215, 217
creativity 5, 11, 17, 18, 38, 42, 49, 78, 98, 140, 150, 177, 185, 203, 210, 225, 229–231
critical theory 102
cynicism 111–113

D

Dark Mountain 14, 80, 119
De Beauvoir, Simone 15, 115, 116
decarbonise 6, 55
deep ecology 49, 104, 178, 179
deliberative democracy 108–111, 164
denial 4, 5, 11, 20, 27, 28, 34, 59, 63, 64, 78, 136, 153, 187, 213, 220, 228
Descartes, Rene 89, 91, 191
despair 2, 7, 9–12, 20, 31, 33, 38, 59, 63, 76, 78, 81, 83, 86–88, 100, 101, 103, 106, 111, 118, 129, 137, 143, 148, 180, 183, 193, 207, 230
diversity 10, 12, 64, 67, 75, 96, 130, 140, 160, 171, 177–180, 206
divestment 43, 54, 130, 219, 228
Donne, John 165
double bind 172
Duggan, Joe 19
dukka 147

E

ecological modernisation 14, 78, 79
ecology 3, 13, 48, 63–65, 67, 102, 115, 124, 126, 128, 133, 137, 139, 149, 167–171, 178, 184, 196, 206, 209, 211
ecosystem 3, 4, 9, 12, 16, 25, 26, 31, 69, 91, 119, 138, 158, 168–170, 177, 182, 205, 206, 237, 242, 243
electric vehicles 29, 203
Ellis, Albert 87
emergency 3–5, 10, 11, 17, 18, 21, 33, 41, 50, 55, 57, 61, 69, 72, 75, 98, 102, 108, 173, 196, 201, 202, 206, 207, 214, 220, 228, 231
emotional resilience 6, 10, 12, 20, 33, 146, 229
Energiewende 12, 50–53
energy demand 4, 31, 158, 213, 228
energy efficiency 4, 29, 31, 51, 203, 213, 221, 228
energy transition 17, 32, 50, 51, 53, 56, 149, 212–217, 219, 221–225
Engels, Frederick 95, 96
Enlightenment 13, 78, 81, 88, 89, 91–93, 95, 191
Epictetus 87
evidence 2–5, 11, 12, 15, 21, 24, 27, 28, 30, 42, 53, 61, 67, 69, 77–80, 82, 85, 86, 95, 96, 104, 105, 110, 117, 127, 131, 132, 144, 168, 172, 174, 176, 184, 186, 195, 199–201, 203, 214, 216, 228–230
existentialism 113
existential risk 4, 5, 21, 26–28, 201
Extinction Rebellion 59, 60, 108, 141, 142, 230
Exxon 186

F

faith 6, 11, 15, 29, 78, 80, 89, 90, 92, 99, 109, 115, 117, 118, 121, 123, 124, 129, 131, 142, 144, 147, 168, 207, 230
feminist 88, 91, 115, 190–192
Figueres, Christiana 146, 202, 203, 206

First Nations 12, 13, 42, 60–63, 73, 76, 142, 229
Fisher, Mark 196
Flanagan, Richard 23
food 13, 23, 34, 65, 75, 84, 95, 106, 110, 132, 135, 142, 144, 149, 156, 184, 202, 203, 205–207, 218, 221, 223, 237
fossil fuel 4, 5, 11, 22, 27, 28, 31, 32, 42–44, 46–49, 54, 55, 59, 64, 79, 82, 90, 102, 105, 113, 120, 132, 140, 143, 146, 186, 203, 213, 214, 217, 219, 224, 236, 237
Frankenstein 94, 95
Frankl, Viktor 100, 111–113

G

Gates, Bill 79
German Coal Commission 52, 53
Gilding, Paul 28
Gill, Gerry 103
globalization 186, 189
gratitude 16, 39, 73–76, 129, 150, 158, 230
Greek philosophy 13
greenhouse gas 2, 4, 22, 30, 54, 85, 108, 140, 182, 235, 236
Greenland 25
Green New Deal 12, 54–56, 220
Gross National Happiness (GNH) 155, 179
Guterres, Antonio 22, 156

H

Habermas, Jurgen 109, 110, 163, 164

Halifax, Joan 157, 158
Hamilton, Clive 118
Hanh, Thich Nhat 146, 148, 150, 157
Haraway, Donna 207
Hayhoe, Katharine 130–132
health 4, 5, 10–12, 20, 21, 25, 31, 33, 60, 72, 79, 112, 132, 142, 155, 156, 200, 205–207, 215, 223, 225, 238, 241, 242
Hegel, George 92
Heidegger, Martin 14, 103–105
holism 169
Holocene 5, 77, 78, 188
honesty 3, 6, 18, 20, 112, 119, 148, 208, 229
hope 1, 2, 7–11, 24, 44, 72, 78, 80, 81, 86, 102, 112, 117, 119–121, 123, 129, 132, 142, 143, 157, 158, 177, 181, 187, 193, 198, 207, 208, 214, 227
Hopkins, Rob 195
Horkheimer, Max 14, 101, 102, 109
hyperobjects 182, 183

I

imagination 18, 41, 44, 59, 86, 193, 195, 201, 203, 225, 231
impermanence 11, 16, 41, 145, 147, 148, 153, 156, 157, 230
Indigenous 6, 11–13, 21, 60–66, 68–72, 75, 76, 79, 88, 116, 191, 202, 205, 229, 230, 235–237
Indra's Net 182
interdependence 11, 16, 39, 40, 60, 70, 71, 127, 128, 130, 148,

153–155, 160, 165, 168, 169, 186, 205, 206, 235
Intergovernmental Panel on Climate Change 21, 130
Islam 138–140
Islamic Declaration on Global Climate Change 16, 140
Israel, Jonathon 89, 92

J

Jack Kornfield, Jack 148
Jemison, N.K. 208
jen 164, 165
Jensen, Derrick 8
Jetnil-Kijiner, Kathy 23
Judaism 134

K

Kant, Immanuel 89, 90, 114, 179, 183
Keats, John 93, 99
Keimowitz, Alison 118
Kingsnorth, Paul 14, 80, 119, 120
Klein, Naomi 116, 200

L

Lamb, Michael 86
Lane, Mary 84
Lane, Melissa 86
Latour, Bruno 16, 167, 185–190
Laudato Si 126, 127, 129
leadership 5, 32, 56, 108, 124, 129, 146, 211, 219, 225, 228
Lear, Jonathon 9
Lent, Jeremy 80, 91
Levitas, Ruth 197, 198

Liberate Tate 12, 47–50
Limits to Growth 174, 197
localist 189
long emergency 3, 7, 16, 33, 184, 207, 208, 230
love 9, 11, 14, 15, 65, 66, 74, 76, 106, 112, 113, 117, 123, 128–132, 135, 149, 150, 152, 157, 160, 172, 183, 230
Loy, David 153

M

Macy, Joanna 8, 9, 148
Mandela, Nelson 9
Marshall, George 23, 29, 44
Marvel, Kate 10
Marx, Karl 95, 96
May, Rollo 9, 104
McKibben, Bill 33, 41, 43, 50, 128, 151
Meadows, Donella 16, 167, 174–177, 180
meaning 3, 6, 9, 15, 61, 81, 91, 100, 101, 103, 104, 112, 114, 116–118, 123, 124, 136, 147, 164, 167, 181, 207, 227
Miéville, China 198
modernity 14, 99, 186
Monbiot, George 119, 120, 197, 213
Morton, Tim 16, 167, 181–185
Moser, Susan 8

N

Naess, Arne 16, 167, 178–180
narcissism 67

nature 7, 14, 15, 39, 40, 48, 68, 88, 91, 92, 95–97, 100–103, 108, 109, 112, 119, 125, 127, 130, 134, 138, 139, 147, 149, 154, 160, 163, 168, 171, 173, 179, 180, 184, 186, 187, 190–192, 206, 237
negative emissions 115, 214, 216, 221, 224
Nietzsche, Friedrich 7
non-human life 178, 179

O

Ocasio-Cortez, Alexandria 54, 56, 112
ocean acidification 25, 156, 201
oil 31, 42, 47, 48, 74, 135, 140, 144, 154, 158, 186, 203, 218, 223, 228
optimism 7, 11, 146, 203
350.org 12, 42, 43, 45, 50, 53
othering 115, 116
Owen, Wilfred 99

P

Pacific Climate Warriors 12, 44–46, 50, 230
Paris Climate Summit 4, 23, 218
Pinker, Stephen 14
Plato 13, 81, 83–85, 91, 191
policy 2, 4, 24, 33, 43, 50, 53, 55, 56, 60, 61, 69, 72, 80, 85, 95, 108, 114, 140, 146, 153–156, 160, 175, 178, 179, 186, 187, 199, 204, 209, 211, 217, 218, 228, 237
Pope Francis 124, 126–128

Potawatomi Nation 69, 70, 73
Potsdam Institute for Climate Impact Research 19, 25, 53
public transport 202, 203
purpose 3, 5, 6, 9, 15, 38, 61, 68, 81, 90, 103, 117, 119, 123, 137, 152, 155, 167, 174, 227, 234

Q

Qur'an 137, 138

R

radical hope 6, 9
Rand, Ayn 151, 152
Raworth, Kate 155, 156
reason 11, 13, 14, 27, 32, 44, 52, 53, 76–80, 85, 87–90, 92, 93, 99, 109, 114, 118, 121, 132, 137, 149, 191–193, 195, 201, 229–231
reciprocity 13, 67, 71, 72, 74, 75, 154, 165, 193, 206, 230, 231
reforestation 203, 209
regenerative farming 228
renewable energy 4, 22, 29, 31, 51, 52, 55, 108, 120, 123, 214, 217, 221, 224, 228
resilience 5, 8, 12, 14, 31, 33, 35, 45, 46, 60, 61, 64, 69, 70, 72, 76, 87, 98, 106, 124, 131, 155, 177, 204, 206, 208–210, 221, 230
Robinson, Kim Stanley 201, 207, 209
Rockström, Johan 53

S

Saint Francis of Assisi 125, 126
Sang-Jin Han 163
Sano, Yeb 22
Satre, Jean-Paul 15, 113, 114
Schellnhuber, Jon 25, 120
Schonfeld, Ralph 83
School Strike 4 Climate 57, 214, 219, 230
Schumacher, E.F. 154
science 1, 19, 32, 39, 73, 76, 78, 89, 99, 121, 130, 131, 149, 154, 168–170, 182, 186, 198, 201, 203, 208, 221, 229–231
seaweed 39, 202
Seneca 87
settler colonisation 70
Shelley, Mary Wollstencraft 93, 94, 99
Shiva, Vandana 191, 193, 206
Shue, Henry 204
smart grids 217, 221
Socrates 13, 81–83
Solnit, Rebecca 8, 37, 207, 208
Spinoza, Baruch 13, 88, 92
Steffan, Alex 28
Stevenson, Wen 105, 106
stoicism 87
Stoknes, Per Espen 7
suffering 9, 11, 16, 17, 63, 64, 95, 102, 116, 118, 130–132, 135, 136, 142, 145, 147, 148, 151, 154, 157, 209, 212, 215, 230
Sunrise Movement 12, 54, 56, 219, 230
Synder, Gary 16

T

Taoism 161, 162
Tao Te Ching 159, 160
techno-fix 173
technological innovation 29, 31, 32, 79, 81, 221, 230
terrestrial 188–190
thankfulness 13, 15, 73, 123, 128, 230, 231
Thucydides 81
Thunberg, Greta 1, 56, 57, 59, 214
Tianxiaweigong 163
tipping points 3, 6, 12, 17, 20, 22, 25, 33, 40, 50, 53, 69, 72, 80, 97, 115, 195, 208, 213, 215, 219, 227
Tirosh-Samuelson, Hava 134
transformational change 12, 17, 35, 37, 97, 174, 175, 179, 195, 198, 220, 221, 225, 228, 230
transition 2, 4, 5, 12, 29, 31, 32, 38, 52, 55, 60, 68, 74, 130, 132, 154, 195, 212, 213, 223, 228, 237
tree planting 202, 203
Trump, Donald 43, 92, 112, 152, 188, 189, 196, 214, 220
trust 13, 48, 60, 71, 72, 86, 136, 186, 187, 193, 196, 208, 214, 224
Tzu, Lao 159, 162

U

United Nations Environment Program (UNEP) 21
Universal Declaration of Rights of Mother Earth 204, 239
utopian 102, 197, 198, 211

W

Wahl, David 206
Wall Kimmerer, Robin 13, 72, 73
Warren, Rachael 26
Waskow, Authur 133
White, Lynn 125, 126
Whyte, Kyle 13, 69–73
Williams, Raymond 7
Windingo 75
wisdom 1, 3, 6, 10, 13, 16, 29, 33, 34, 40, 60, 61, 63, 73, 78, 81–83, 85, 86, 117, 118, 121, 124, 126, 130, 135, 145, 147, 150, 159, 207, 209, 214, 225, 227, 229, 231
Wisdom in Nature 139

Woolf, Virginia 100
Wordsworth, William 93
World Meteorological Organization 21
Wright, Erik Olin 198–200

Y

Yunkaporta, Tyson 13, 66–69

Z

zero-carbon economy 5, 29, 38, 60, 74, 98, 154, 204, 212, 224, 228
zero emissions 4
Zinn, Howard 37, 39

GPSR Compliance
The European Union's (EU) General Product Safety Regulation (GPSR) is a set of rules that requires consumer products to be safe and our obligations to ensure this.

If you have any concerns about our products, you can contact us on

ProductSafety@springernature.com

In case Publisher is established outside the EU, the EU authorized representative is:

Springer Nature Customer Service Center GmbH
Europaplatz 3
69115 Heidelberg, Germany

www.ingramcontent.com/pod-product-compliance
Lightning Source LLC
LaVergne TN
LVHW040734250326
834688LV00031B/281